Factory Farming

Factory Farming

ANDREW JOHNSON

BLACKWELL
Oxford UK & Cambridge USA

First published 1991

Basil Blackwell Ltd
108 Cowley Road, Oxford, OX4 1JF, UK

Basil Blackwell, Inc.
3 Cambridge Center
Cambridge, Massachusetts 02142, USA

British Library Cataloguing in Publication Data

A CIP catalogue record for this book is available from the British Library.

Library of Congress Cataloging in Publication Data

Johnson, Andrew.
Factory farming/Andrew Johnson.
p. cm.
Includes bibliographical references and index.
ISBN 0-631-17843-0
1. Livestock factories. I. Title.
SF140.L58J64 1991
363.19′2 – dc20 90-47620 CIP

Typeset in 10 on 12 pt Garamond
by Graphicraft Typesetters Ltd., Hong Kong
Printed in Great Britain by T. J. Press Ltd., Padstow Cornwall

Contents

Acknowledgements

I should like to thank the following people who have provided material or helpful criticism:

Stewart Angus, Professor Donald Broom, Richard Cottrell MEP, Winnie Ewing MEP, Dr David Flint, Dr Michael Fox, Paul Jackson, Jim Hill, Dr Marthe Kiley-Worthington, Dr Roger Mitchell, Ann Petch, Dr James Serpell, Professor John Webster.

The generous assistance of the following voluntary organizations is also acknowledged:

Chickens' Lib, Deutscher Tierschutzbund eV, Foreningen Til Dyrenes Beskyttelse i Danmark, Greek Animal Welfare Fund, Liga Portuguesa dos Direitos do Animal, Royal Society for the Prevention of Cruelty to Animals.

Particular thanks are due to Peter Roberts and the staff of Compassion in World Farming for their advice, assistance and encouragement. Also to my wife Alison who kindled an interest in animal welfare, and to Mary Midgley whose arguments convinced me that animals really do matter.

Part I

MAN THE CONTROLLER

1
The Agricultural Enterprise

For most of us, the word 'farm' evokes an image of a plain but dignified dwelling set in a patterned landscape of neatly tended fields; green with peacefully grazing flocks, brown with clean new furrows, or golden with the rippling wealth of harvest. Add to this picture one or two haystacks, a few hens and perhaps a duckpond, and you have the quintessential setting for rustic life. The scene is described economically by the dictionary definition: 'farm, n. a tract of land used for cultivation and pasturage, along with a house and other necessary buildings'; it is pictured in rhapsodic hues on the wrappings of 'golden farm' eggs, 'Olde English' hams and 'traditional mature' cheeses, and in subtler tones on the tiled murals of bijou delicatessens or country fayre restaurants.

But if you travel a hundred miles along one of Britain's motorways, which afford a view of our historic farmlands relatively unimpeded by coaching inns, straggling villages and the other detritus which flanks long-established highways, what you will see from the road is rather different from the picture I have just drawn. The patchwork of fields has largely been replaced by endless tracts of monoculture: as far as the eye can see there is nothing but wheat, or sugar beet, or Friesan cattle. Haystacks and hedgerows have been replaced by concrete silos and chain-link fences, and pretty buildings have given way to flimsy shacks of asbestos or corrugated iron, some venting wisps of foul-smelling steam. And what is more, in all that hundred miles, it is unlikely you will spot a single pig or chicken, although in Britain's supermarkets their union-jack bedecked joints and carcasses line the shelves in their millions. What has happened to farming, and where have all the livestock gone?

Ask this question of a growing number of people concerned about animals' rights and the environment, and you will hear a gruesome story. The animals have all been put into prisons, though their only crime was that in the open air they took too long fattening for slaughter

A typical intensive livestock unit. The numerous ventilation ducts through the walls and roof are characteristic of buildings designed to house large numbers of animals per unit area. (Compassion in World Farming)

to please their greedy masters. The descendants of yesterday's free-ranging herds are now subjected to horrific indignities in the squalor of dark insanitary sheds, where they are often confined so closely that they can neither turn round nor lie down. No longer do farmers care for their stock on an individual basis, but they have relegated the animals to the status of cogs in a ghastly machine, whose sole purpose is to make money. Quite apart from the cruelty of modern agriculture, any friend of the earth will tell you of its disastrous impact on the environment we live in, and on the flavour and wholesomeness of the food we eat.

Put the same question, on the other hand, to one of the so-called prison wardens, and you will get a very different answer. The animals, it seems, are safe and warm, well-fed and happy in their houses – if this were not so they would not yield so well. Their masters are unfairly abused public servants, without whose ceaseless toil we should be threatened by hunger and privation. And the pastoral myth is just that: the reality of life on the open range was cold, short and miserable.

Which answer should we accept? Is everything in the garden lovely, or is there really something rotten in the state of agriculture – something that cries out for reform? In part, it depends on whether we regard the efficient, cheap production of food as an end of such priority that farmers are justified in putting all welfare and environmental issues very firmly in second place, or whether we take a more holistic view which demands consideration of these latter questions as soon as we have enough to satisfy our basic needs. But before we can even discuss such issues, it seems prudent to survey the concrete facts of modern farm husbandry, and how it differs from traditional methods. Even here, there are sharp disagreements between the farmers' critics and their supporters. On the one side, the past appears as a golden age of harmony between man and beast, while on the other it is characterized by famine and savagery. It might seem pointless to worry about such historical details, since whether we wish to or not, we are not in a position simply to turn back the clock; but they come up so frequently in the debate that they cannot be ignored. Besides, some knowledge of the better and worse aspects of past practice can be very useful in helping to assess the possibilities for future reform.

The history of farming is inextricably linked with the history of civilization itself, for before humans learnt to practise agriculture in the vicinity of their settlements, a nomadic and culturally primitive life style was more or less inevitable. It is widely presumed that for many thousands of years before the last Ice Age, our distant ancestors lived in small itinerant groups, hunting and gathering over a wide area for a variety of plant and animal foods. In favoured areas, such as the sheltered plains south of the Taurus and Zagros mountains, it is possible that small permanent villages subsisted on the pickings from their immediate surroundings; and archaeological evidence suggests that it was in this 'fertile crescent' that people first learnt to cultivate and harvest wheat and barley, and to store the grain for times of shortage. This agricultural way of life encouraged permanent settlements such as Jericho, probably the world's first city, which was founded some 10,000 years ago.[1]

As soon as cultivation of crops became a significant part of life, attitudes to property must inevitably have changed. The hunter–gatherer must move on whenever food becomes scarce, so all but the smallest personal possessions would be too much of a burden. And if he does not shoot the antelope or gather the berries he passes today, what matter if someone else does? When nineteenth-century anthropologists began to study the few remaining tribes of hunter–gatherers, they were astonished to discover how little they worried about the future. Rather than storing food against possible shortages, they behaved 'as if the game were locked up in a stable', taking only what they could use immediately. But the farmer who has cleared the ground, sown the seed he stored last autumn, and weeded the growing crops, has done all this work in anticipation of the reward of harvest time. He will feel a strong attachment to his piece of ground, and will be inclined to stand and defend it against raiders. By virtue of his labour, he claims as his own property both crops and land.

The origins of animal husbandry are even more speculative than those of crop growing, since there is less archaeological evidence for the transition from hunting to keeping. But it is reasonable to suppose a progression from merely following wild herds to warding off predators, then either driving the animals to new pastures, or reserving certain areas of ground for grazing, or possibly even growing crops specifically for the livestock. Again, with each stage of domestication there is an increasing input of labour in the interest of ensuring future plenty, and an ever-greater incentive to regard the animals as property to be owned and traded in.

By appropriating to himself control of which crops grew on his land and where his flocks could graze, the early farmer could produce a virtually guaranteed output of food beyond his own needs and those of his household. The first city dwellers, and their successors ever since, have depended on this surplus: in cultures where it could not be sustained, either from the homelands or from the farms of conquered territories, decline has been inevitable. Except in areas where seasonal flooding supplied fresh nutrients to the soil each year, it must have soon become apparent that a piece of land could not successfully grow the same crop year after year. The Old Testament book of Leviticus stipulated that the land should be left fallow one year in seven, though it seems doubtful whether such a brief respite would have helped very much.[2] Just before the birth of Christ, the Roman poet Virgil was much more detailed in his advice:

> Though flax exhausts the soil and likewise oats
> And poppies drowsy with Lethean dreams,
> The land can bear them in alternate years
> If you are generous with rich manure
> And scatter ashes on the worn-out fields.[3]

When the Romans came to Britain, they found the natives alternating grain crops with fallow year about, and they introduced a three-year rotation (wheat/beans/fallow), which was still in use in the eighteenth century. Since then, many other variations have been tried, of which the best known is probably the four-year Norfolk cycle of barley, grass and clover, wheat and turnips. Under this system, the alternation of crops and grazing animals maintained fertility and also ensured control over weeds, crop diseases and pests.

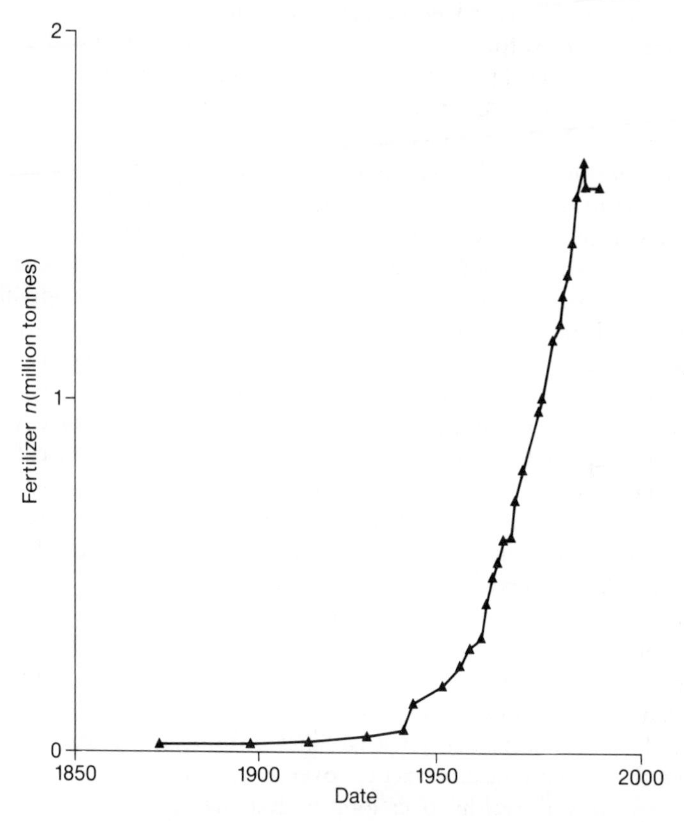

Figure 1 Fertilizer use in the United Kingdom.
(*Source*: London Food Commission *Food Adulteration* p. 116.)

The rotation of crops which make different demands on the soil is an important aspect of conservation, but will not necessarily guarantee continued fertility. As crops, wool, dairy produce and meat are sold off the farm, there will be a gradual depletion of nutrients from the soil. In some soils, natural weathering of mineral particles can replace these losses, but very often deficiencies develop which will reduce yields unless the missing elements can be replaced by applying animal manure or other fertilizers. Ploughing and harrowing may also be responsible for the loss of nutrients from the soil, as they can greatly accelerate leaching out by rainwater. The problem of declining yields is still of major importance, although in the more advanced agricultural nations it is often disguised by the application of ever-increasing amounts of artificial fertilizers. (Figure 1 shows the increased use of fertilizers in the

United Kingdom.) Virgin land just cleared of its natural vegetation may grow excellent crops for a year or so, but unless the soil can adjust to a new equilibrium, cropping will not be sustained. Unfortunately, our considerable technical knowledge has not stopped us degrading the newly cropped lands of America, and overtaxing the long-established farms of Europe, for short-term profit at the expense of permanent stability. Contrast the carefully improved plots of the Maoris in New Zealand, whose efforts created a deep black soil which has remained stable even in places where it has not been worked for over a century.

Far worse than loss of fertility caused by unregulated cropping or grazing, which time may perchance repair, is the wholesale loss of soil by erosion which so often accompanies the farming enterprise. The natural landscape in many areas shows a considerable degree of resistance to the forces of wind and water, due to the existence of covering vegetation adapted to its native habitat. When such vegetation is burnt, cut down or overgrazed, the finer soil particles may be washed or blown away, leaving behind coarse and unfertile sand or even bare rock. Ancient civilizations in the Mediterranean thus reduced fertile lands to 'goat deserts', and currently the burning of tropical forests to clear land for grazing is having the same disastrous effect. When land is cultivated for cropping and bare soil exposed to the elements, there is a very high risk of erosion – particularly in windy areas, or where the ground has more than a slight slope. It is possible to take preventive measures, as witnessed by the spectacular terraced hillsides of ancient Peru, but to be effective they need concerted action over a wide area.

To keep the soil healthy over long periods has always been a major problem for farmers, particularly when population growth or other economic factors have put pressure on them to increase production at all costs. Modern agricultural technology and artificial fertilizers can ameliorate some of the effects of soil degradation, but it can well be argued that increasing dependence on high-tech methods is a reckless course which is merely storing up trouble for the future. We are familiar with reports of ecological disasters in developing countries, where agricultural technology is sometimes applied inappropriately in an effort to make up for lost time; but it should be remembered that one of this century's worst environmental disasters occurred in the world's most advanced agricultural economy, when farmers tried to switch back from ploughing the prairies to grazing cattle, and created the American Dust Bowl.

Even if he can master the soil, the farmer may still find it difficult to profit from his plants and animals. The first species he chose to domesticate were well enough adapted to their accustomed environment by natural selection, but this does not mean they inevitably fared better in their new artificial surroundings. And with animals in particular, breeding from a limited captive stock offers no guarantee of improved adaptation in subsequent generations, but rather tends to lead to more weakly and degenerate offspring. Although the farmer tries to provide the conditions for his crops and livestock to flourish, he cannot give them complete protection from predators and disease, but must watch in anguish as locusts devour or unkent fevers wrack. Such problems, too, are still with us for all the armoury of pesticides and drugs that modern science has made available. It has been estimated that while pesticide use in the United States increased ten times over a thirty-year period, crop losses from insects actually increased during the same time.[4] Looking at figures for the whole world, over 30 per cent of all crops grown are lost to pests of one sort or another before harvest.[5]

Despite the fact that human beings have been farming for more than 10,000 years, it is evident that many of the problems faced by farmers remain just the same: first, the preservation of the soil, on which all else depends, and then protecting and nurturing crops and livestock until the time of harvest or slaughter. But is it possible to go further, and use the historical approach to find some ethical common ground as to how livestock are treated? At first sight such a project looks most unpromising: the attitudes of earlier cultures towards other animals are heterogeneous, and frequently bizarre. Pliny reports a tribe who have a dog for their king, 'and him they obey, according to the signs which he maketh by moving the parts of his body',[6] while countless primitive people worship animal deities, a practice continued by the Egyptians long after their hunting and gathering days were finished. For the Romans, fierceness was the characteristic animal trait, and the populace revelled in such spectacles as the massacre in a single day of 200 lions, as many leopards and 300 bears.[7] The Jews were, as usual, much concerned with the ritual uncleanness of numerous species, while for the witch-hunters of Tudor England almost any small animal was suspected of being an evil spirit in disguise. And in all times and places there have been many people who kept particular animals which they regarded quite simply as friends.[8] Where are the lessons from history in such a jumble of customs?

While the attitudes of distant civilizations are often strangely incomprehensible, the colourful examples just cited do not necessarily reflect the more humdrum concerns of those who actually worked the land. Their preoccupation with the constant challenges of agriculture would inevitably result in attitudes far less changeable than the whims of the non-farming members of the community. For the farmer, his livestock are first and foremost his property, to be fed and cared for now because of their future value. This matter-of-fact relationship does not preclude either a tender and caring attitude to the beasts, or a callous and lazy brand of husbandry, but it does set limits beyond which the farming enterprise cannot prosper. It also implies that the degree of care given to different animals will depend quite a lot on their future usefulness, witness the attention given to young animals from birth to maturity, and the frequent neglect of those too old for breeding or milking. Before tractors took over their duties, elderly working horses and oxen were particularly vulnerable to ill treatment as their usefulness declined.

There are practical limitations to the ways in which a responsible farmer should treat his livestock, and it can be argued that sheer prudence demands that he neither neglect nor abuse them, since to follow either course is to risk losing the reward of all his efforts. However, this specious argument, which is vigorously promoted in Britain by the Ministry of Agriculture and the National Farmers' Union, is only partly true. Against it, the claims may be pressed with considerable force that modern farming practices are *not* responsible in the long term, and that merely keeping animals alive in a state of efficient productivity can often involve gross and unnecessary suffering. More particularly, that no matter how kind or harsh the farmer, many of the ways in which animals are now bred and housed are inherently more cruel than the methods of fifty years ago. Such a counter-claim need not attack the whole farming enterprise, but can be urged quite coherently from within the tradition described above. It is, of course, possible to argue that if we were all vegetarians there would be no livestock to suffer (though the ecological consequences of such an unlikely transformation are debatable), but a less radical challenge to farming methods has the considerable advantage of starting from where we're actually at, and stands a far greater chance of achieving majority support for actual reforms.

The common purpose of agriculture today with that of previous ages both defines the farmer's problem and sets limits to its solution. He requires a fair return for his labour, but will only get this in the short term if his husbandry is reasonably efficient, and in the long term if he preserves or improves the fertility of the land. Unlike the hunter-gatherer, who accepts the opportunity and challenge of untamed nature with little thought for the future, even the most primitive farmer aims at a degree of control over nature, in the hope of ensuring a full larder throughout the year. This aim has not changed, only the means by which control is achieved, from the simple digging stick of Mesopotamia to the modern combine harvester, or from the methods of biblical shepherds to those of today's broiler merchants.

Any livestock farming involves providing the animals which are intended to produce eggs, milk, wool or meat with an environment in which production can be achieved satisfactorily. Despite the modern obsession with efficiency, measured simply in terms of market value versus costs, this need not be the only criterion for choosing between different methods of husbandry. Other factors include overall environmental effects, welfare consideration, a whole variety of assessments of the quality of the produce, and religious fetishes (particularly regarding slaughter).

As well as his immediate concern with the end-product, the farmer needs to replace beasts which die or are slaughtered, usually by allowing some of his own animals to breed, very often with others bought in specially for this purpose. His aims in breeding livestock can equally simplistically be reduced to sheer profitability in terms of future gains and present costs, but in the real world there is fortunately still some concern for welfare, as well as an important aesthetic tradition regarding what a 'good' animal should look like.

Judgement of what farm animals should be like logically precedes considerations of how they should be kept, though its welfare implications are at first sight less obvious. In past centuries, uncontrolled breeding in small flocks or herds might lead at worst to a random proportion of weakly offspring with congenital abnormalities. Modern selective breeding allows quite precise choice of various characters useful to the farmer, yet has been allowed as a side-effect to produce cows that cannot calve except by Caesarean section, poultry whose rapid growth strains legs and lungs to the limit, and pigs so clumsy they cannot look after their own offspring. Further ethical questions are raised by the widespread use of artificial insemination, and the more recent techniques of embryo transplants and genetic manipulation, which may soon be established parts of farm practice.

Whether by accident or by design, farmed species are bound to become different from their wild ancestors. Domesticated wheat, for example, has the seeds more firmly attached to the rachis or upper stem, because cultivation has favoured an uncommon mutant variety of the wild plant with a tough rachis. The seeds of the tough rachis strain are less easily lost during harvesting, thus the amount in cultivated crops will inevitably increase. When the increased proportion became noticeable, possibly after a thousand years or so of cultivation, farmers may have deliberately begun to select the better yielding type for next year's seed.[9] The choice of livestock for domestication must also have produced changes, as traits with survival value in the natural environment became less important, and other characteristics were favoured instead. Farm animals would inevitably become easier to catch and more docile in confinement, and it is probable that they have always been deliberately selected on such grounds as size and generally healthy appearance.

It is difficult to say at what stage in the history of farming control of breeding became important, but by the earliest historical times it is clear that there was artificial selection of which animals bred together. By only allowing certain pairs of beasts to mate, the farmer can systematically influence the way in which his herd develops. Although the formal theory of genetics is very recent, a number of classical authors give rules of thumb for choosing suitable sires and dams, and they were obviously aware of the way in which characteristics might be inherited from grandparents while missing the intervening generation.[10]

Modern breeds of farm animals differ considerably from those of even 200 years ago. In Britain, the enclosures of the eighteenth century, and the introduction of fodder crops such as turnips, meant that the hardiness and small appetites needed by animals kept on common land were no longer so important. By breeding from the best examples of local stock, and recording in detail his successes and failures, Robert Bakewell made spectacular improvements in the size and quality of both cattle and sheep. His methods were widely followed, and many present-day breeds date from this period of improvement. The breeding strategies of Bakewell and his followers involved careful selection on the basis of appearance and pedigree, and relied on the general axiom that 'like begets like'. Breeders this century have the advantage of a more detailed knowledge of the principles of heredity, enabling them to solve such puzzles as why the offspring of two roan coloured cattle are often plain red or white instead of resembling either parent.[11]

Despite scientific advance, successful breeding remains a difficult art, in which shrewdness is required in asking the right questions, as well as in finding the answers. It is relatively simple to change average size over a period of time, as shown by experiments in which growth rate and size at maturity of mice were doubled by fifteen generations of selective breeding;[12] but while many breeding programmes have aimed without question at rearing larger animals, such a strategy is not necessarily profitable. The advantages of size vary from species to species: with pigs and poultry, larger breeds slaughtered earlier are economic, because younger animals put on more weight for a given amount of feeding. Against this improvement, however, must be set the greater expense of maintaining a parent population of oversized animals.[13] When the offspring being reared for slaughter account for the major part of food costs, as is the case with pigs or poultry, it pays to rear large breeds and kill them young; but this advantage does not apply with cattle, where food costs are equally distributed between supporting the mother cow and fattening the calf, while for sheep the balance is actually in favour of smaller breeds, since the ewe requires more sustenance than the lambs she can bear, and bigger ewes will eat more extra than is recouped from the sale of bigger lambs.

Even if the primary objective of a breeding policy can be defined, as for example minimizing overall feeding costs per kilo of usable carcass weight, there are other factors which must be considered even on purely economic grounds. In the discussion of the advantages of fast-growing breeds of pig or fowl, the quality of the product was not mentioned. In fact, the meat is leaner, which appeals to those worried about their intake of fats, but it lacks flavour, and often has a repulsively gelatinous texture. Marketing, and the price advantages involved, have largely overcome consumer resistance to such tasteless pieces of flab; but in the beef industry, insistence by butchers on a certain level of maturity before slaughter tips the balance in favour of Hereford and Angus cattle against the larger Charolais and Limousin. As a general rule which affects profitability more significantly, the accentuation of particular traits, by inbreeding among the animals in which they are favoured, tends to result in a decline in general hardiness and an increase in congenital abnormalities. This problem may partly be overcome by breeding from parents of two different pure strains to give offspring which show 'hybrid vigour', but even so there is much disturbing evidence that so-called improved breeds suffer from higher rates of disease and mortality than their less developed ancestors. Such effects are only too familiar in many overbred varieties of dogs and cats, as is the gross exaggeration of certain physiological features by breeding, exemplified by the pug's squashed nose or the Cavalier spaniel's popped eyes. While not directly painful, such developments as the rearing of turkeys carrying so much breast muscle that they cannot mate naturally, or of Belgian Blue cattle whose freak double muscling means they cannot calve except by Caesarean section, raise ethical problems of their own. Furthermore, a potentially dangerous gap is opened between the ability to survive in anything like natural conditions, and dependence on the specialist life support systems of modern intensive husbandry.

The stakes in this risky game have been raised dramatically by the increasing standardization of breeds, the loss of such potentially useful types as the Lincolnshire Curly Coat pig or the Suffolk Dun cow, or the

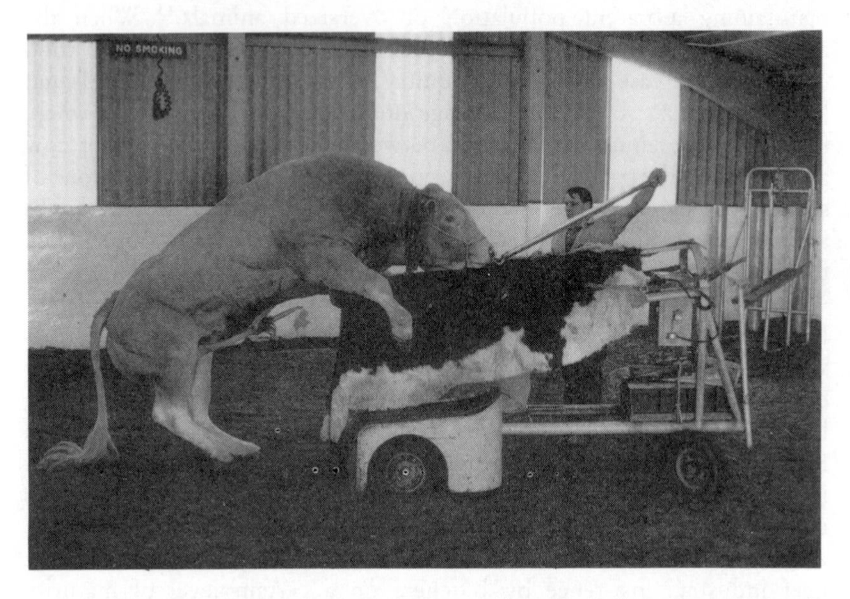

A cow-skin covered framework is used to collect the bull's semen for use in artificial insemination of cows. It is now possible for one bull to father many thousands of offspring. (© Patrick Sutherland)

twenty-one other British breeds which have become extinct since the start of this century; and even more so by the ever-greater dependence on artificial insemination. When one bull can sire over quarter of a million offspring, the consequent loss of genetic diversity is enormous. Even if the first generation of calves are healthy enough, further inbreeding among them may show up hitherto unsuspected defects. The effects of a restricted gene pool were illustrated in the United States when inbreeding among Jersey cattle caused the build-up of a recessive gene which causes a crippling deformity known as 'limber leg'.[14] On the female side of the pedigree, each mother has till now made her own contribution to the genotype of her offspring, but this may not be the case much longer. A technique has recently been developed in which eggs removed from a chosen and specially killed cow are fertilized in a test tube, sexed, and the embryos implanted in the womb of a recipient cow. By such an immaculate conception a dairy cow could 'mother' twin beef calves. One only wonders what the rest of the herd would make of the miraculous event – perhaps it is just as well they can't tell us. Embryo splitting or more advanced cloning techniques can be used to produce large numbers of identical siblings, or two embryos can even be fused to create chimeras such as sheep with goats' heads. There does

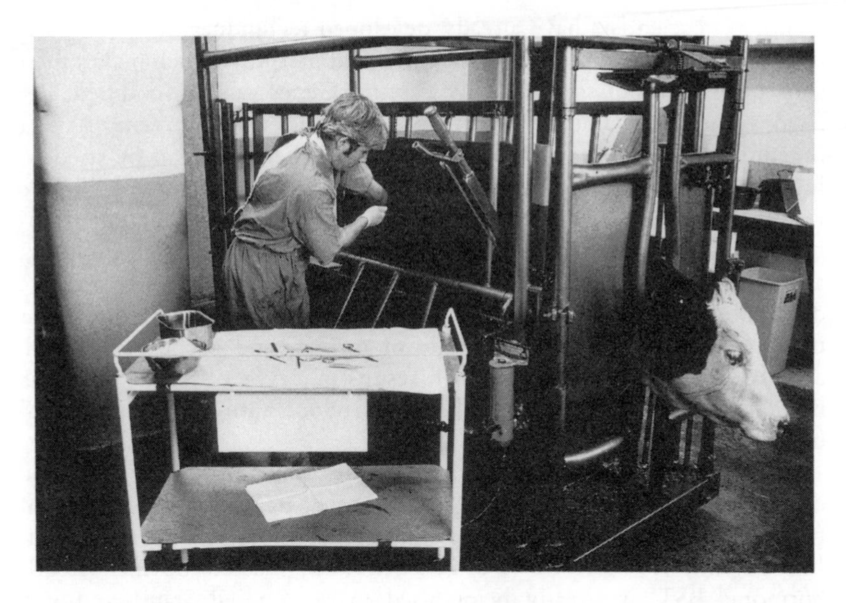

The number of calves produced from a single pedigree female can be increased by transplanting embryos into carrier cows, which act as surrogate mothers. (© Patrick Sutherland)

seem to be a risk, though, that while developments like this may give impressive results in the short term, if their use becomes widespread there may be unexpected and unwelcome consequences in the future.

This caveat applies with even greater force to the novel form of microbiological manipulation which has been dubbed 'genetic engineering'. All the inherited information the embryo requires to determine its subsequent development and the characteristics it will have as an adult (environmental factors being equal) is carried on chromosomes in the cell nucleus, which consist of DNA molecules derived partly from the mother and partly from the father. In normal reproduction, though each parent effectively contributes just a dozen or so chromosomes to the offspring, pieces of information (genes) which start out on one chromosome frequently cross over on to another, so that the genetic information passed on to the embryo is pretty well random. Some genes which are very close to each other in the molecular structure of the chromosome do tend to go together, and are said to be linked.

Although there is a very long way to go before we can tell the exact contributions of each section of DNA to the inherited characteristics of the embryo – a task compounded in difficulty by the fact that the contribution of each gene is often modified by the presence or absence of others – scientists have already developed techniques in which parts of the DNA can be split off and recombined in a way which makes their subsequent transmission virtually certain. Genes can be modified, deleted, cloned, or manufactured in large quantities using fermenting yeast or cultures of bacteria. Recombinant DNA chemistry can be used to insert genes from completely different species, frequently with bizarre results. For example, a tobacco plant has been produced which incorporates genetic material from a firefly so that its leaves glow in the dark, and mice which have been given human growth genes grow to rat-like dimensions. More constructively, genetic engineering has already been used to manufacture a purer grade of insulin, and a cheaper vaccine against hepatitis B. The gene which results in a type of haemophilia has been transferred to sheep, which then produce antibodies for medical use.

As far as the 'creation' of new species goes, a major interest in genetic engineering at present is in the production of new viruses as potential pesticides, but in the long term new food animals may well be produced. In conventional agriculture, a genetically engineered bovine growth hormone, BST, is already being used in the United States to boost milk yields, and trials have been carried out in Britain, where they have given rise to some controversy. The use of similar hormones for pigs or broiler chickens could lead to faster growth rates, and it has been suggested that genetic engineering could adapt species to bad climates or confined housing, or improve their disease resistance. Further applications proposed include the production of crops resistant to herbicides to simplify weed killing among growing crops, and the elimination of genetically transmitted human diseases by a programme of screening and genetic manipulation.

Apart from the ethical implications of such techniques, concern has been expressed about the wisdom of incorporating genetic material from one species into another. The production of viruses which may affect creatures other than their intended targets, or which may mutate to cause unforseeable damage, is a risk that is very difficult to assess. Attitudes towards the release of genetically engineered organisms into the environment vary: in Britain, the official line is fairly relaxed, and several planned releases of viruses have already taken place, while there has been intense controversy in the United States, and several European countries have legislated to prevent such occurrences. Transgenic animals can now be patented in the United States, so that for seventeen years royalties must be paid to the 'inventors' on all their progeny. This development has been attacked by the majority of animal welfare organizations on the ground that such patenting provides an extra economic incentive to research which is likely to be inimical to ethical and animal welfare considerations. Genetic engineering also carries the risk, as with traditional selective breeding, that useful characteristics may be lost from the gene pool altogether, and that herds which lack genetic diversity may suffer catastrophically from unforeseen diseases.

So much for down the line financial and prudential concerns regarding the way we control the breeding of farm livestock. In economic terms, market price less costs is all that matters, and so long as morbidity from overbreeding does not push the vet's bills too high or make an impact on yields, it can be put up with. And if risks cannot be quantified, economists have a great way of ignoring them. Fortunately, the real world is not yet quite like this. There are still many farmers who take a pride in the overall appearance and state of health of their flocks, and who count this reward of good husbandry above the smile their bank manager gives when he reads the bottom line. There are even, in my part of the world, quaint old farmers who like their cows to be served by a real bull, on the assumption that they enjoy his attentions more than those of the artificial insemination man. And there may somewhere, in a yet more primitive state of development, be peasants who actually imagine their beasts might fancy one mate rather than another, and who do not grudge them the choice.

It is easy to dismiss the attitudes of different times and places as absurd and simple-minded, and indeed we cannot help but see through the eyes of our own generation, accepting, by and large, the commonplace and rejecting the unfamiliar. But we should realize that we may have our own blind spots, too, and the laboratory boffin or city speculator is just as capable of faulty judgement as the person with mud on his boots. To take just one example, the question of optimal size discussed above, both academics and hard-nosed businessmen frequently demonstrate emotional *naïveté* as great as that of the oystercatcher which can't resist a fake egg nearly as large as itself. Genetic engineers all too often assume that faster growing animals will necessarily be more efficient, while entrepreneurs enthuse over such prospects as 'a dairy cow as big as an elephant and capable of producing 45,000 lbs a year'.[15]

To ask, on a grand scale, what are the rights and wrongs of controlled mating or other interference with the natural reproductive process is as unlikely to produce a generally acceptable answer as questioning the ethics of keeping farm animals at all. But this is no objection to considering the welfare implications of different methods now in use, and seeking agreement over more limited reforms. It can certainly be urged that a technique which involves unnecessary pain (that is, which causes more pain than some other way of achieving the same objectives) is undesirable, and the majority of people would agree with this even if there were some slight penalty in cost or convenience.

Whatever methods of husbandry are used, the human factor must always be considered. Good stockmanship can make a major contribution to animal welfare and productivity, and is very often more important than the physical aspects of a particular system, but as long as this is borne in mind, there is no objection to comparing different techniques on the assumption of reasonably competent operators.

Without trying to compile any sort of farmyard *Kama Sutra*, it may be as well to go into slightly more detail about what respectable farm animals regard, or have become accustomed to regard, as normal sex. As the slightest acquaintance with domestic or wild animals shows, copulation in most species is preceded by elaborate courtship rituals in which the two sexes alternately encourage and ignore each other. To watch the attempts of a ram or bull to charm the partner of his fancy, and her coy but unconvincing refusals, is as amusing as it is instructive. Mating occurs only when the female is on heat, during a limited period of the oestrus cycle; the length of the cycle being around two to three weeks for most farm animals, and the period of heat varies from twelve to twenty-four hours in a cow to one or two days in a ewe or up to three days in a sow. Though it is easily discernible to conspecifics, the stockman requires considerable expertise to recognize oestrus: signs to watch for are females mounting one another, general excitability, and a slight discharge of mucus. Pigs and cattle will mate and breed at any time of year, but sheep normally mate only in the autumn, thus producing lambs the next spring.

Rams are normally turned out with the ewes around a particular date, and left to get on with it; and beef bulls likewise. It is possible to record copulation of animals in the field by attaching a harness to the male's chin or chest, from which coloured ink or chalk is transferred to the female during mating, but it may well be the case that a male running free will cover some females many times and others not at all. A common practice with dairy cattle and pigs is 'in-hand' covering, where the female is restrained and the male helped up on her if need be. The advantage is that each pair couple once and only once, but against this apparent improvement in efficiency, there is evidence to suggest that familiarization with males increases the fertility of females.[16] And it is known that as a cow (or indeed a human) becomes sexually excited, uterine contractions and secretions occur, which will assist the passage of sperm up the vagina to fertilize the egg. The absence of preparatory courting can also, therefore, be expected to reduce conception rates.

Quite apart from the practicalities of achieving successful fertilization with in-hand service, there are possible objections to the practice on welfare grounds. Since under natural conditions oestral females only stand for the male after prolonged courtship and stimulation, it may be supposed that the abrupt mounting of a tethered female could be traumatic from her point of view. Evidence for this point of view is provided by Kiley-Worthington's report that maiden mares covered in-hand without preliminary courtship show subsequent signs of trauma such as shivering, sweating and avoiding the area where they were 'raped'.[17] Such objections are difficult to evaluate or discuss rationally because long usage has accustomed us to prefer quite different expressions when speaking about sex in humans or in other animals, exemplified by the terms 'rape' and 'in-hand service'. To many people the former has an unacceptable emotional content, while to others the latter is an equally unpalatable euphemism. Scientific studies of physiology and behaviour in our own and other species have to some extent narrowed the gap, but the ideal of an entirely objective terminology is practically impossible to achieve. However, efforts in this direction have stimulated debate about animal welfare considerably, not just regarding sex, but over the whole spectrum of activities.

Natural copulation is not without its dangers, particularly when the male is heavy or elderly. One further argument for in-hand mating is that it allows the use of service crates, which relieve the female of some weight, and reduce the risks of injury by crushing or scratching. Such dangers are very real in certain breeds of pig, which have been bred for exaggerated carcass growth at the expense of viability under natural conditions. The more modern alternative of artificial insemination, already widespread for cattle and becoming increasingly prevalent among pigs, provides an alternative method of fertilization which is more reliable and more humane than old-fashioned in-hand service. Artificial insemination provides the farmer with a relatively cheap way of improving his herd, particularly for the small breeder who lacks the capital to buy a top class bull or boar, although as a national policy it is subject to the caveats already mentioned concerning genetic diversity. If the scheme is properly regulated, the incidence of venereal diseases can also be reduced. Compared with natural mating, artificial insemination is labour intensive, and requires skill from the stockman in detecting when the females in his charge are on heat. (A recent American development can overcome the latter problem: a tiny thermometer implanted near the vaginal opening of a cow signals the progress of her oestral cycle to a computer.) The same disadvantages also apply to in-hand mating; but while artificial insemination is no doubt much less pleasant than the joys of untrammelled sex, the indignities suffered by the recipient animal are arguably much less than they often are where traditional forced coupling is still the rule.

Breeding replacement stock involves more than just mating of the intended parents. The conditions allowed for pregnancy, birth and subsequent maternal care of the offspring will influence the success of a breeding programme both in terms of economics and of animal welfare. However, these conditions vary so much from species to species, and are so closely related to the ways in which non-breeding stock are housed and fed, that it will be more convenient to consider them as part of a more general discussion of the confinement, housing and feeding of livestock, in which each of the common types of farm animal is dealt with separately.

2

From Farm to Factory

Farmers have always been concerned with boundaries. Among tribes with no written culture, anthropologists have often remarked on the taboo signs attached to boundary posts or fences as a curse against trespassers, while the landmarks of the ancient Babylonians and Greeks were charged with inscriptions of similar intent. Indeed, the English custom of 'beating the bounds' of a parish was traditionally accompanied by curses on those who transgressed their neighbours' bounds, and blessings for those who respected them. The very word 'farm' is derived from the 'firma' or fixed payment which entitled the occupant to possession of his defined tract of land. Occupation involves responsibility towards neighbours as well as rights against them, and unless the farmer was a powerful landlord, in which case he could grant himself the right for his pigeons, pheasants or deer to maraud the tenants' crops with impunity, it must always have been an important duty to prevent his livestock from straying and causing damage. English law makes it clear that it is in general the stock owner's responsibility to keep his animals in, rather than his neighbour's duty to keep them out.

From ancient times, the farmer has had two choices available: he can keep a more or less constant watch over his beasts to stop them straying, or he can erect a physical barrier for the same purpose. In most of Britain since the great enclosures of the eighteenth century, the herdsman's job so defined has virtually ceased to exist, though large areas of Scotland are still grazed by sheep under a regime where the beasts' movements are controlled only partly by fences, and to a greater extent by periodic shepherding of the traditional variety and reliance on the natural tendency of the flock to remain more or less together in one area. In the South, however, the animals are no longer driven off crop land to more distant pastures in the summer and brought back home after the harvest, and Little Boy Blue is no longer needed to keep the cows and sheep in their respective places: his job is now done by permanent walls and hedges, or by movable electric fences.

For grazing animals, confinement in fields does not mean a completely unnatural or overcrowded environment. In terms of efficient production, there is no point putting more beasts on a given area than the grass can support. (In the distorted terms of subsidy agriculture it may make sense: where I live, a starving ewe has more value than no ewe at all.) If fields are used for prolonged grazing, parasites will inevitably be a problem, to be dealt with either by frequent movement of the stock or, more often, by regularly dosing them against worms and flukes. Since the amount of grass available varies enormously according to the time of year, growing hay and other fodder crops for winter feeding will considerably increase the number of livestock that can be kept on a given acreage. If the animals are being kept in near natural conditions, access to water and some shelter from wind or sun are their main requirements beyond adequate supplies of food. Since we no longer have wolves or grizzly bears in Britain, the general need of livestock to be protected from predation hardly applies to our larger farm animals such as sheep and cattle, though new-born lambs still need protection from foxes and crows.

The scope for keeping pigs or poultry in anything like natural conditions is very limited on most farms. The pig is a creature of the forest, but as long as it is given shelter from the sun and palatable food to substitute for its natural diet of roots and acorns, it can survive under a wide range of conditions. Likewise the hen, far removed from its native jungle, needs only food, water and protection from predators such as fox or mink to keep it laying eggs for gentlemen. It may be said that the adaptability of these creatures is their greatest misfortune. Try keeping a wild goose or an ostrich in a cage for eggs, and even if it doesn't die of boredom you'll get very few eggs in the span of a year. Try fattening hippopotami for the Sunday roast and you may enjoy a juicy fillet or so, but to breed a supply of baby hippos plentiful enough to replace those slaughtered will be a practical impossibility. But the poor old hen goes on laying, the sow bearing more piglets, for better or worse, in sickness and in health, till death's final release. It doesn't show they're happy, any more than a woman's periods show she's happy; it's just the way they're made.

Cattle and sheep are consumers of pasture, and keeping them is a way of profiting from grassland which is not rich enough ground to bear other crops, or which may be cropping land at present lying fallow. So until very recently, when this truism has sometimes been lost sight of and grazing beasts are seen only as producers of milk, wool or meat, and not as consumers of grass, sheep and most cows have been inseparable from their pastures, except maybe for some temporary winter shelter.

Even in ancient times the oxen used for working were habitually kept inside, as sometimes were bulls, to keep them away from the cows. The practice of fattening calves or lambs for the table in 'little dark cabins' was current in the seventeenth century, and was no doubt known about long before, but it was presumably for a specialist luxury market. But while pigs will weed broken ground, and hens will keep grass down in an orchard, this is not why they are kept, nor do they need such habitats to perform their vital task for the farmer – the conversion of cheap fodder, which he provides, into high grade animal protein that he can sell at a profit.

It is idle to suppose that in the past this inevitably meant they were free to ramble among the buttercups in some odd corner of the farm, where they were fed occasionally and otherwise left to get on with it. This picture can indeed be true, and a few pigs and poultry fit very nicely into such a niche on the old-fashioned mixed farm; but Varro describes much more specialized Roman chicken farms, with elaborate hen-houses equipped with ladders, high roosts, nests and reliable trap-doors to keep out foxes and weasels. These held from 40 to 200 birds; the larger houses being subdivided into several rooms so that each cock and his attached hens could roost apart from other 'families'. Hygiene was maintained by piping smoke from the bakery through the hen-house in order to destroy lice and mites.[1] Periodic evacuation followed by disinfection of the building to curb the build-up of diseases and parasites is just as essential today in any building where livestock are permanently housed.

The sixteenth-century pig, according to Keith Thomas, lived in considerably less comfort than the chickens on such a model Roman farm:

> In Elizabethan times the usual way of 'brawning' pigs was to keep them 'in so close a room that they cannot turn themselves round about . . . whereby they are forced always to lie on their bellies'. ('After he is brawned for your turn', the formula continued, 'thrust a knife into one of his flanks and let him run with it till he die; [or] gently bait him with muzzled dogs'.)[2]

'Sweat-boxes' of exactly this type are still in use today, though not very common in Britain. (Slaughter methods, at least, have become rather more humane.) And the modern battery house is little more than a many thousand times larger replica of the housewife's kitchen hen-coop which might at that date have filled in the unused space under the dresser.

The ever-increasing scale of modern livestock houses is part of their nightmare quality for their detractors, and doubtless part of their appeal for their admirers. Up to 100,000 chickens under one roof is quite common, with automatic feeding and watering. For laying birds, labour requirements are reckoned to be one stockman to 30,000 birds; in a broiler unit of this size just one person might be employed to check them over occasionally. The development of such large units is relatively recent; in 1961 97 per cent of UK laying hens were in flocks of less than 1000, while by 1983 only 12 per cent were in flocks of less than 5000, with 43 per cent kept in units of 50,000 and over.[3] This represents the culmination of a movement from labour-intensive to capital-intensive farming which has been going on for the past 200 years.

The success of the industrial revolution depended on the increased efficiency of large scale production (based on mechanization and more effective division of labour) and on the possibility of trading over greater distances. The increased costs of supplies and distribution had to be covered by lower production costs if the new methods were to be profitable. The long period that elapsed before the application of industrial methods to agriculture must have been due at least in part to apparently insurmountable problems of transportation. With small farms, animals could graze and forage wherever they were herded, and the only necessary movement of fodder was bringing in the winter stores to conveniently situated barns. To transport such low value commodities as hay or turnips over long distances would have been prohibitively expensive. Distribution of farm produce to customers was also costly, though increasing urbanization meant that more large scale movements of food were to some extent inevitable. Even so, the greater the distance involved, the worse the impact on profit margins, particularly for perishable goods where time spent in transport is bound to mean increased wastage.

In the past, the most efficient way of distributing meat to the butchers' shops was to walk the livestock to market, where they were sold on the hoof. This could be done over substantial distances, as exemplified by the trade in Scottish beef cattle, which were being driven to markets in the South of England as early as the twelfth century.[4] While cattle and sheep could be moved around the country in this way, it would be less practical with pigs, and impossible with poultry. Regional specialization in farming of the two latter, at least on a large scale, was thus a less attractive proposition. Dairy farming, too, was practised in all areas, because of the need to get the milk to the consumer in a very short time. Some parts of the country were particularly involved in cheese production from surplus milk, but the scope for such specialization was limited, because these surpluses are more seasonal than local. In cities, cows used to be kept in byres all the year round: as recently as

1928 there were over 100 such herds within the boundaries of the city of Edinburgh.[5] Freshly calved cows were imported from England, kept for up to two years, and slaughtered when their milk yield dropped.

Until the present century, large scale animal husbandry was virtually unknown. Of course, some landlords had immense holdings, and made or spent fortunes on them, but the scale of the actual farming operations carried on by their tenants was limited, and small mixed farms were very much the rule. A catalyst for change in Britain was the development of a nationwide rail network in the nineteenth century, which made cheap and fast transport available to all areas. Such perishable goods as soft fruit, flowers, and of course milk, could be at market in London within four or five hours of being delivered to a station in far away Cornwall. Increased specialization, and a trend towards the economies of scale offered by larger units, were the inevitable results as competition from farmers in areas more favourable for particular crops or livestock hit those in less favoured regions. This trend has continued right up to the present day, with cattle and sheep production being favoured in the wetter western half of Britain, while intensive pig and poultry units tend to be situated in the east, where the cereal crops used to feed them are grown.

About 100 years ago, the use of refrigerated ships opened up the meat trade to international competition, with regular cargoes from Australia and the Argentine seriously affecting the viability of the long-established Scottish producers. Other methods of preservation such as canning and deep freezing have also had a significant impact on farming, as the operators of such capital-intensive plant have sought to ensure the supplies they need either by going into farming themselves or by negotiating long term contracts with specialist producers.

More specialization was accompanied by more mechanization, as crop growers invested in tractors and combine harvesters, dairy farmers installed milking machines, and everywhere employers sought to reduce escalating wage bills by installing any device which was guaranteed to save labour. Moving livestock around the farm in the traditional way is time-consuming work, which demands certain skills. In the modern factory farm, this problem is neatly dodged by keeping the stock in one place and installing highly mechanized or automatic systems to convey their food, water and waste-products. Since large scale intensive livestock units were introduced in the 1920s, they have come to dominate the British market in egg and pig meat production, so that 90 per cent of laying hens and a third of breeding sows are now kept in close confinement. This means they live most or all of their lives indoors in cages or stalls which are hardly larger than the animals themselves. In fact, the

The wing-span of a chicken is about twice the width available in a typical battery cage. It would be an offence to keep a pet bird in such cramped conditions of confinement. (Compassion in World Farming)

wingspan of a laying hen is about 75 cm, yet in a normal battery system a typical cage for *five* hens is only 45 × 50 cm. One animal welfare group has related this space allowance to human experience by claiming, 'the equivalent would be for you to spend your life in a wardrobe – with two other people'.[6]

The opponents of factory farming insist that it necessarily involves cruelty to the animals: Chickens' Lib point out that anyone keeping hens under battery conditions would probably have been prosecuted seventy-five years ago (under the then recently passed Protection of Animals Act of 1911), yet now it is regarded as normal practice. However, supporters of intensive methods point to the higher price of produce from other systems of husbandry, and claim that the livestock do not suffer unduly. Some, of course, go further and make extravagant claims about land shortages or food shortages which would inevitably result if factory farming were to be abandoned; these are in fact completely without justification, except possibly as irrational reactions to the more extreme elements of the Animal Liberation Front. It cannot be denied that shop prices favour the produce of at least some intensive systems, and if it is to be argued that any of these systems be banned, then reasons must be produced which outweigh the price increases that such a course would inevitably bring. Such reasons may be justified by selfish concern about food quality, altruistic concern for animal welfare, or general concern for the environment at large. Before examining the evidence on any of these counts, however, it seems appropriate to examine more closely some of the current practices for intensive husbandry of various species, with an eye to their advantages rather than their defects.

The modern battery egg unit is the acme of intensive livestock housing, in terms of efficiency of production and of its almost universal adoption in the industrialized nations. In both Britain and the United States, over 90 per cent of the laying flocks are in battery cages: some 35 million birds in the United Kingdom and 300 million in the United States. Russia has even more layers than the United States, while the Japanese keep some 150 million chickens in cages. The dominance of the battery system is relatively recent; although the technology of wire mesh production is over 100 years old, and various patterns of battery cages were being patented in the 1920s, less than 20 per cent of British eggs came from batteries in 1961. At that date, deep litter houses accounted for about 50 per cent of egg production, with the remaining 30 per cent free range.[7] The post-war development of antibiotics encouraged more intensive production, as did the connection of farms to the national electricity grid, which enabled farmers to exploit the advantages of artificial light as a stimulus to laying, electric fans to cool buildings in hot weather, and so on. The number of laying chickens in one building is typically around 30,000, with British and European farmers allowing about 450 sq cm of floor space per bird, while in the United States the recommended area is as little as 315 sq cm, or a little less than a single page of this book. Birds in individual cages lay less well and require more space, so cages normally hold four to six birds, and are arranged in tiers along opposite walls of a long building. The sloping floors of the cages ensure that eggs roll forward on to a rack or conveyor belt as soon as they are laid, while another belt distributes food, and droppings either fall into a pit under the cages, or are removed by yet another moving belt.

An important advantage of battery houses for the farmer is the close control he can exercise over various environmental factors to optimize the number of eggs laid annually by each hen. Temperature can be maintained at 21° C, and daylight extended to seventeen hours right through the winter, when birds kept under less intensive regimes lay fewer eggs. The use of artificial lighting involves some extra energy costs, but these can be reduced by using an intermittent light programme, with illumination for just 15 minutes of every hour during the birds' day. It is even possible to arrange a twenty-eight-hour cycle of lighting so that the hens experience a six-day week, and lay fewer, but larger, eggs.

Under normal management, about 250 eggs per hen per year can easily be achieved. While this is about double the production rate of 1945, much of the improvement can be attributed to the development of better rations and selective breeding to increase the number of eggs laid: comparable figures for free range and deep litter birds are now of the order of 200 to 225 eggs per year, and one recent report actually showed that over an eighty-week period, brown egg layers on range produced more and bigger eggs than the same strain of birds in cages.[8] In the battery unit, however, the birds do not need to eat as much to keep warm in cold weather, and their rations can be portioned out more effectively, with less chance of greedy birds bullying the more timid feeders. And it is claimed that disease and mortality rates of battery hens are generally lower than those of hens kept under other systems, due to the more hygienic environment in which droppings are removed from the birds as soon as they are produced. Battery houses also score in terms of convenience to the operator, by the reduced labour requirements, the arrangement of cages to allow rapid checking of all the birds and the removal of any that are dead or dying, and the cleanness and ease of collection of the eggs. The capital cost per bird is less than for most other systems on account of the enormous number of birds that can be crammed into a large battery unit; and with lower labour costs as well, the economic advantage of the battery system is such that the eggs can be produced at significantly lower cost than those from less intensively kept hens. This does not necessarily mean higher profits to the farmer though. In fact, at present, the premium paid for free range eggs is due in a large part to the excess of demand over supply, and it has been estimated that free range eggs can be over three times as profitable to the farmer.[9] By and large, however, the people who have made big money out of eggs in the past forty years have been the battery boys rather than the traditionalists.

For all its attractions to the commercially-minded, who may care little about the quality of the product or the environmental impact of their operations, and even less about the sensibilities of their livestock, the battery system has certain operating disadvantages which must be overcome to make it profitable. Although such parasitic ailments as coccidiosis have been eliminated, because the hens do not have contact with each other's droppings or with contaminated earth or litter, many other diseases thrive in the warm and crowded conditions of the battery. Even if production levels are maintained, egg quality may suffer if the hens are unfit. Birds with infectious bronchitis, for example, produce bleached, misshapen and thin shelled eggs with watery whites.[10] Infectious bronchitis, Marek's disease and fowl pest can all be vaccinated against, but without complete success, and the potential for loss is considerable in a building containing 20,000 or so birds. There are also many non-infectious complaints to which the battery hen is particularly susceptible. Lack of exercise can be blamed for fatty degeneration of the liver, followed by haemorrhage and sudden death; and for a common complaint known as 'cage layer fatigue', which is a general skeletal weakness that may also be aggravated by dietary deficiencies.[11] Since battery hens cannot peck around to supplement their feed, it is vital that it should contain all the necessary trace elements. Even the physical characteristics are important; it has been shown that severe mouth ulcers are widespread in hens fed on an all-mash diet, which does not contain solid particles of the size needed to stimulate normal mouth functions.[12] The crowded conditions of the battery may also encourage hens to attack each other, though cannibalism, feather-pecking and vent-pecking are not unknown in other husbandry systems, and over a third of all layers have the ends of their beaks removed in an effort to overcome these 'vices'. In the confinement of the battery cage, of course, the victim has less chance of escape from this sort of aggression, nor will it necessarily be observed by the stockman, who can only easily see the front ends of his hens. The layout of the battery can make it difficult to spot a number of other ailments, as illustrated by this report by Chickens' Lib on an egg-bound bird:

> '[The] bird was affected with egg material in an abnormal site associated with the pressure of a fully formed egg in the oviduct. At some later date this would almost certainly have developed into an egg peritonitis.' The hen in question had deteriorated to this condition and the stockman had not noticed because her entire abdominal region was hidden by the food trough etc. of the battery cage. She was in a cage on the top tier, and only her head, neck and feet were visible to any stockman.[13]

To overcome the above problems, a number of therapeutic measures are adopted. Viral infections can be controlled by antibiotics, which can also be used to boost egg yields. Routine administration of virginiamycin is claimed to improve feed consumption per egg laid by 4 or 5 per cent, while bacitracin stimulates egg production, and oxytetracycline improves eggshell quality. It has been estimated that about 50 per cent of British layers are dosed routinely with antibiotics; in the United

States the figure is nearly 100 per cent. Subdued lighting can help avoid cannibalism, but may also make inspection more difficult; and as with any system of husbandry, competent and conscientious supervision of the flock is an important factor in determining overall efficiency.

Laying hens are most productive in their first year of lay, from eighteen or twenty weeks old to about seventy-six weeks. After this, it may be worth keeping the birds for a second year since although fewer eggs will be laid they will be larger, or the hens may be sent straight for slaughter as 'spent layers', to be used for making soups, chicken pies and baby foods. If they are being kept for a second year of laying, it is in the farmer's interest that the birds get through their unproductive moult period as quickly as possible. Forced moulting, in which the flock is given no food, water or light for twenty-four hours, followed by two weeks on a low protein diet, with a gradual return to normal lighting, was officially forbidden in Britain by the Welfare of Battery Hens Regulations 1987, but it is not clear to what extent the ban is effective. In the United States forced moults are the rule if a second year of laying is required, with birds being denied water for up to ten days.[14]

The spent layers which go to slaughter at specialized poultry processing plants are only a small portion of their business. The plump, pink, plastic-wrapped birds, whole, portioned or ready-cooked, which brim forth from chill cabinets and freezers in every supermarket and corner shop, are very different creatures. While the hybrid layer has been bred for maximum egg production, the object of the broiler chicken's life is quite simply to convert fodder efficiently into tender juicy hunks of meat. As already discussed, this aim is achieved by breeding for rapid growth and killing young; so broiler chickens are off to the slaughterhouse long before their battery cousins have even started to lay, usually at the age of six or seven weeks. By this time they weigh around 2 kg (4–5 lb), compared with a weight at the same age for a chick of a normal laying strain of around 450 g (1 lb). The increase in growth rate that has been achieved through selective breeding and routine doses of antibiotics is quite remarkable: twenty-five years ago broiler fowls did not reach the slaughter weight of 4½ lb until they were fourteen weeks old.[15] As a result of this progress, shop prices have fallen and there has been a dramatic increase in the consumption of chicken in Britain, from 50 million birds in 1955 to approximately 500 million in 1988.

Virtually all chickens for meat are reared in broiler houses, in which light and temperature are maintained at the optimum levels for promoting growth, and the floor is covered with an absorbent layer of wood chips or straw. The birds are allowed about 450 sq cm each, which gives them plenty of room to move about when they are small, but they

In a modern broiler shed, animal welfare is critically dependent on the close control of environmental factors such as ventilation and the state of the floor litter. Under such crowded conditions control of infectious diseases is a major problem; the majority of chicken carcasses are infected with salmonella, and chicken meat is responsible for a high proportion of food poisoning incidents. (Paul James)

become crowded as they grow bigger. The sheds are brightly lit when the chickens are small, to encourage them to move about and find food and water, but as they grow larger the lights are dimmed to discourage aggression. Debeaking may be carried out, particularly with birds that are intended for the Christmas trade, and have to be kept to an older age in order to grow to the required size. Illumination for twenty-three and a half hours a day is normal, the half hour of darkness simulating a power cut, which might otherwise result in panic, with many birds being injured or suffocated. Food and water are supplied automatically, so the main job of the stockman is to remove dead birds and check on the general condition of the flock.

The chickens from one unit are all sent for slaughter together. Catching and packing into crates is often done at night to avoid panicking the birds by exposure to daylight. At the slaughterhouse, they are shackled

Slaughter arrangements for poultry are highly mechanized. This turkey, still fully conscious, has just been shackled on to the production line. From this point the process of slaughter, plucking and packing is fully automatic. (© Patrick Sutherland)

by the feet to a conveyor which takes them to an electric stunning bath, followed by an automatic killer which cuts their throats. The carcasses then proceed to a scald bath before plucking, cleaning and packing.

The degree of automation, highly bred stock and controlled environment all contribute to the impressive 'productivity' of the modern broiler industry. Meat quality, however, leaves much to be desired when compared with the traditional farmyard chicken. The taste of any food animal's flesh is inevitably affected by what it has been eating itself, and if the broiler chicken's diet contains fish meal and bits of dung-soaked litter, it can hardly be regarded as surprising if its meat is tainted with a certain *je ne sais quoi*. Its high rate of growth and lack of exercise must also contribute to the broiler's lack of gourmet appeal, by making its flesh distinctly soft and flabby. While profitability demands high stocking rates and the use of ultra-rapid growing strains, both of these factors can cause health and welfare problems. Overcrowding may lead to feather-pecking and cannibalism unless illumination is kept very low, and the birds may be severely stressed by the difficulties they face in getting access to feeders and drinkers. Sudden deaths are common, particularly in hot weather, and mortality rates have increased as productivity has risen.[16]

The most common causes of death are colisepticaemia, which may be associated with vaccination against such diseases as infectious bronchitis, and fatty liver and kidney syndrome similar to that reported in battery layers.[17] There are a further range of non-fatal conditions which can result in down-grading of carcasses and reduced prices. Their rapid growth rate and lack of exercise makes broilers particularly prone to leg and foot disorders, and older birds spend most of the time sitting with their weight supported by the upper legs and the breast. If the litter is hard and wet, blisters and dark coloured ammonia burns disfigure the flesh of hocks and breast, and make the trussed bird look unappetizing. In severe cases, hock burns must be scraped back to the bone, since the bacteria they harbour make the damaged areas unfit for human consumption. Bruising, as the birds are being caught or earlier, also causes discolouration, so that parts of the bird may have to be rejected, and the rest used for chicken joints. However, despite these negative points, the broiler industry has been a huge commercial success, creating a vastly increased market for chicken meat, and while free range alternatives are beginning to reappear, so far they have made very little impression on the broiler monopoly.

Turkeys are reared by similar methods to broiler fowls, and their production has expanded in the same way, from three million birds per annum to nearly thirty million in the past twenty-five years.[18] Stocking densities in turkey sheds are rather lower than for chickens, and instead of sending the entire flock for slaughter at once it is normal to cull the female birds progressively at thirteen, fifteen and eighteen weeks for the oven-ready market, while males are grown on to twenty-four weeks before being converted into turkey ham or turkey sausages. Ducks can be grown under similar conditions, but goose farming remains a more traditional activity. Geese only breed during the spring, and their eggs are more difficult to hatch than chicken or turkey eggs. From about three weeks of age the goslings are allowed outside to graze during the day, and most of them are slaughtered at Michaelmas or Christmas. In 1988 nearly 500,000 geese were sold for Christmas – four times more than in 1982.[19] The perception that geese are a more natural product than most turkeys is probably responsible for this increase in demand.

Geese on the continent do not lead such natural lives however. In France alone, nearly two million birds a year are force fed to produce *foie gras* for gourmets. Until they are four months old the geese range freely and feed on grass, but for the final three to four weeks of their lives they are made to eat vast quantities of cooked maize, and deprived of exercise, so that while the bird itself increases in weight by about 60 per cent, the liver swells to four or five times its previous weight. Traditionally, the maize mash was forced manually down a funnel into the bird's throat, and exercise might be prevented by nailing its feet to the floor. In more modern establishments, an Archimedean-screw type machine is used to stuff food into the birds, which are kept in small cages between feeds. Elastic bands may be put around their necks to prevent food being retched up. Mortality during *gavage* is up to 10 per cent, from such causes as lesions in the gizzard and intestines, heart failure or liver necrosis.[20]

The virtually total displacement of old-fashioned ways of poultry keeping by intensive undercover techniques has not been followed in the farming of most other species of livestock, for which a wider range of husbandry methods are in current use. But in Britain and the United States more than half of the 100 million or so pigs eaten every year are bred from sows kept inside in close confinement systems. The advantages of keeping pigs indoors are similar to those for poultry: heat, light, food and water can be precisely regulated, and in a concrete floored shed the farmer does not need to worry about damage to pasture by rooting and trampling. Once it has been decided to keep the herd under cover, economics favour the highest possible stocking density, which may be conveniently achieved by an array of individual stalls. Needless to say, the wire mesh used for chicken batteries would not restrain a hefty pig, so their cages are made of stout steel tubing, reminiscent of prison bars.

The breeding sow spends the sixteen weeks of her pregnancy in a stall two feet wide, with a floor which may be all concrete, or partly slatted. In some stalls the pig is entirely enclosed by bars, while other patterns are open at the back, in which case she is tethered by a chain round the neck or a band around the girth. In any event, the sow can stand or lie down, but not turn round. Shortly before giving birth the sow is moved to a pen of different design, known as a farrowing crate, where there is space for the piglets to retreat to when they are not suckling. This separate 'creep' area for the piglets reduces the chance of a clumsy sow lying on her offspring and crushing them, and gives the pig-keeper access to the piglets without having to face the unwelcome attentions of a protective sow. Natural weaning would take around two to three months, but piglets are sometimes removed from their mothers at about three weeks old, and kept for the next fortnight or so in fully enclosed heated cages before they go on to the next stage of the fattening process. The lactating sow is dried up by withdrawal of food and water for twenty-four hours, and at this stage she may be allowed to run in a yard for five days with other sows and a boar. As soon as she is served, the sow is returned to the confinement of the dry sow stall, and the cycle is repeated. If tether stalls are used, in which there are no back rails, it is not necessary to leave the sow at liberty for mating. Instead, she may be returned to the stall immediately after weaning; the boar being taken on a daily perambulation along the line of tethered sows, which allows him to copulate with any that are on heat.

The young growing pigs are usually kept indoors, where they can most efficiently convert feed into lean meat without putting on layers of fat to keep them warm. They may be given straw, but slatted floors are often preferred, to prevent the pigs having access to their own dung and urine. Overcrowding can lead to fighting, with tails particularly vulnerable to attack. Many farmers try to avoid injuries by removing the sharp points of the piglets' teeth, and docking part of their tails. Dim lighting and high stocking densities discourage exercise, and the pigs put on weight rapidly. Some producers still use 'sweat boxes' in which ventilation is deliberately restricted to keep temperature and humidity up, and the animals stand or lie in their own excrement. Pigs are slaughtered at between four and six months old. For fresh pork, a slaughter weight of about 70 kg is preferred, while pigs for bacon, sausages and pies are grown on to around 90 kg live weight.

As with intensive poultry systems, the indoor rearing of pigs affords

Many pregnant sows are kept in cages to simplify their management and avoid the possibility of fighting. The stress induced by permanent confinement produces characteristic behaviour such as bar chewing (left) or head drooping (right). (Compassion in World Farming)

opportunities for the rapid spread of infection, and stringent hygiene precautions are necessary to keep disease at bay. Swine vesicular disease, Aujesky's disease, transmissible gastro-enteritis and streptococcal meningitis are just some of the problems facing the pig industry, and routine use of antibiotics is prevalent. At present, though, indoor systems mean cheaper meat, and competition from Danish and Dutch producers (who operate highly intensive systems) has drastically reduced the amount of outdoor pig farming in Britain, to a small number of breeding herds and one or two small specialist producers of free range pig meat. Indoor methods do vary considerably from those described above, to systems where the pigs have much more freedom of movement in strawed pens or yards.

For many people, factory farming is epitomized by the crated veal calf. While hens can be stigmatized as witless, and pigs as naturally dirty, who can resist the pathetic appeal of the mild brown eyes of a little calf? And the eyes and face are just about all there is to see in the average white veal unit, where the calves are constrained by coffin-like crates, from which their heads stick out towards the feed buckets at meal times. Although the veal crate is illegal now in Britain, it is in general use in the United States and most of Europe, and restaurant veal is almost invariably from crated calves. In fact, much European veal is actually from British calves, of which we annually export several hundred thousand, the majority of which end up in continental crates. To the farmer, the main advantage of the process is the efficient conversion of feeding stuff to meat, under the influence of the same factors of warmth, darkness and lack of exercise as have already been discussed in the context of broiler fowls and sweat-box pigs.

Veal calves are separated from their mothers in the first few days of their lives, and put into the slatted wooden crates where they will spend the rest of their lives. Instead of fibrous food, which they would normally take from age about two weeks, the calves are fed on an entirely liquid diet based on skimmed milk powder, and deliberately formulated to be deficient in iron. The low-iron diet and restricted physical movement of the calves are responsible for the 'white' flesh, which is prized by gourmets. Not surprisingly, this regime is associated with various ailments, in particular pneumonia and diarrhoea, and mortality rates of up to 20 per cent have been reported. A study conducted by the United States Department of Agriculture found that calves in individual crates or stalls required five times as much medication as those kept in outside hutches with yards, as well as showing significantly more lameness and symptoms of chronic stress.[21] But consumer preference for white meat has meant that, apart from in Britain where veal is not a traditional part

These coffin-like crates for veal calves are now illegal in the UK, but many British calves are still exported for rearing in this type of system. The calves are kept in near darkness and on a low-iron diet so that they develop anaemia, producing the 'white' meat favoured by the majority of consumers. (Compassion in World Farming)

of the diet, the close confinement method of rearing continues to be used.

Dairy and beef cattle are still on the whole creatures of the open pasture in most European countries, although they have often been taken inside in the winter months to reduce feeding costs and to avoid the damage to grassland caused by heavy trampling under wet conditions. Covered barns or yards with plentiful straw provided the normal winter shelter. However, increasing specialization has meant that little straw is now produced in some areas, and the high cost of buying it in from other regions, together with the labour involved in distributing and mucking out straw bedding, have encouraged farmers to change to wooden slats or bare concrete. In some areas, herds are now being kept permanently off the fields, which are used to grow silage crops instead. The cattle are then fed on a year-round diet of silage and bought-in concentrates. This sort of zero-grazing husbandry is already widely used in the United States, in association with a variety of intensive confinement systems reminiscent of the old-fashioned town byres described

Slatted flooring reduces the labour costs involved in mucking out traditional straw bedding. But slats are less comfortable for the cattle, and the semi-liquid slurry produced can be a serious pollution hazard. (Author)

above. Many dairy cattle are kept indoors in free-stall barns or tethered continuously in individual stalls, while others are left between milking in outdoor corrals, which may be crowded and muddy, and are often without shade or shelter. Beef cattle are frequently kept in vast 'feedlots' under conditions which would strike most European farmers as unacceptably squalid. Mud and excrement cover the ground, and shelter from heat or cold is often completely lacking. Under these conditions, productivity per animal is likely to be much worse than in situations where individual beasts get more careful attention, but reduced labour costs and economies of scale in purchasing, transport and administration combine to make the intensive systems more profitable.

High stocking densities encourage disease, with cattle as with any other species. The operator of a beef feedlot can shrug off many ailments that may afflict his herd, so long as they do not affect the fattening up process; but dairy cattle must be kept more healthy if milk yields are to be maintained. Although as herd sizes steadily increase, more British dairy farmers are adopting the practice of zero-grazing and intensive confinement, it is not a great success in terms of animal health. Milk production has been increased considerably, using carefully formulated diets of silage, soya beans, fishmeal, dried poultry manure, etc., but such rich feeding puts considerable stress on the cow. The fermentation of grass which occurs in the cow's rumen depends on natural sugars which support microbial activity; in silage making these sugars are converted to lactic acid and acetic acid, which are not only unappetizing, but can cause a condition known as rumen acidosis, resulting in discomfort and lameness. Wet food such as silage also makes the cows excrete far more water, and means it is more difficult to keep them clean and dry. Mastitis, or inflammation of the teats, is a frequent result. The cure, as usual, involves treatment with yet more antibiotics, such as erythromycin or penicillin.[22] A new alternative to silage is the 'maxgrass' process developed by British Petroleum which preserves the grass by using a mixture of formic acid and other chemicals instead of the bacterial fermentation involved in silage making; but what it does to the milk is anyone's guess.[23]

Of the traditional farm animals, that only leaves the sheep. For reasons already explained, it is a highly unsuitable animal for intensive production, although extremely effective as a means of utilizing poor quality grassland. This has not stopped some people trying undercover sheep rearing, but so far without commercial success. An experimental farm for lamb production in Russia suffered a severe blow when two-thirds of the 30,000 sheep died from diseases which spread out of control in the crowded sheds.[24] Broiler lambs remain a theoretical possibility, but with world lamb prices as they are at present, intensive sheep farming is very unlikely to pay, with the possible exception of undercover 'finishing' of fat lambs.

At this point, it seems only fair to say, after such a catalogue of frequently unpleasant practices, that close confinement of livestock under intensive systems is far from being the only subject for concern in the farming world. Even in the realm of animal welfare, many traditional practices are highly objectionable, and nowhere is this more true than in the case of sheep farming. Although they are not subjected to the abuse of the battery, both the vast Australian wool flocks and Britain's subsidized hill sheep offer many examples of sheer neglect, rough handling and primitive home surgery. And however they are kept while alive, very few farm animals meet a natural death, or more to the point a quiet and painless death. Spent battery hens and worn out free range layers alike may end their lives swinging upside down from the conveyor that leads to the stunner, the knife and the scald bath, or being sold at 50 pence from a freezing flea-market stall to be taken home and ritually slaughtered. Large scale, small scale, traditional, modern, are not in themselves terms of approbation or abuse, nor do they necessarily indicate which practices are more acceptable from the point of view of the consumer, the livestock or the environment. But the trend to specialized intensive livestock units has shifted the balance of welfare concern away from the individual and towards the system; with the result that however much a good stockman could help his beasts in the past, the physical realities of modern husbandry methods may severely limit his ability to improve their present lot.

3
Taming and Ownership

Factory farming, as outlined above, displays a recognizable and even logical continuity from the traditional husbandry of domesticated animals. It is true that confinement has become closer and more restricting, that specialization has replaced diversity, and that selective breeding has produced livestock which would scarcely be viable outside their closely regulated environments. The modern broiler fowl, for example, has been so intensively bred for rapid weight gain that the females which are kept for breeding are unable to survive healthily, if at all, unless their food and water intake is strictly rationed. But in all the cases so far discussed, the animals are representatives of species which have been domesticated successfully for thousands of years. Stone Age man kept pigs, and the Babylonians of 5000 years ago kept chickens which were similar in appearance to those of today. Why it is that just these species, and only a few others, have been kept for so long is a matter for speculation; certainly the common farm animals are easy to catch, and not over-inclined to wander off if food is provided at home, but whatever the justification for tradition, recent years have seen a proliferation of new species being farmed for the first time. This trend has been helped by greater knowledge of the food, shelter and sex-life needed by different species in captivity, such information often coming from the numerous zoos established in the past century or so; and experimentation with new species has become very much simpler with the ready availability of cheap wire cages, and the use of temperature-controlled buildings for housing them. Nowadays, if you want to breed mice, foxes, quail, mink – or just rabbits, the solution is the same: shove a few of them into battery cages, provide food and water, and you're away.

Just as Stone Age man evidently found it convenient to keep pigs and cattle at home rather than relying on the luck of the hunt, so twentieth-century man, faced with dwindling wilderness areas and diminishing wild stocks of fur and game animals, has adopted the practice of domesticating creatures which are of commercial value. This must seem like very good sense to the business interests which are ravaging the remaining untamed regions of the globe, since having domesticated the species they regard as useful, the others can safely be allowed to go to hell. That is, of course, just so long as ideas about what is useful and what isn't can be guaranteed not to change. The general reaction to this proliferation of farming of erstwhile wild species has been mixed: for some people it is a logical utilization of resources that are there for our benefit, while for others the caging of wild creatures is an affront to nature, and almost blasphemous. But if we have no right to domesticate new species, what right save that of custom do we have over those already in farms? As Hobbes remarks, 'If we have dominion over sheep or oxen, we exercise it not as dominion, but as hostility; for we keep them only to labour, and to be killed and devoured by us; so that lions and bears would be as good masters to them as we are.'[1]

Whatever the explanation, whatever the motives, there is money in the big business of fur farming: a battery unit of 10,000 mink could produce pelts worth around a quarter of a million pounds each year. In a typical British mink farm, the animals are kept in wire cages about two feet long by a foot wide, and are fed on a diet of fish offal, chicken heads, etc., left on the roofs of the cages for them to pull through. At about eight months old the mink are killed by an injection of barbiturate, and then skinned. The pelts are cleaned and dried, while the bodies are sent for processing into soap or animal feedstuff.

Only about 1 per cent of the annual world production of around twenty-five million mink pelts comes from British farms. The lion's share of this type of farming belongs to the Scandinavian countries, with Denmark and Finland each producing over four million skins, a figure equalled by the United States, while Russian output is a little under three million pelts. The other animal commonly farmed for its fur is the fox, of which some two million a year are bred in Scandinavia – about 75 per cent of the whole world production.[2] Husbandry methods are fairly standard, with very few farms now allowing the animals access to outside runs, though cage sizes and the number of beasts to the cage vary. Many fur farms buy in animal food from specialized factories, normally in the form of a thick paste compounded of slaughterhouse offal, fish waste, potatoes, etc. Even whale meat is sometimes used. At 250 g a day for every mink, it takes over three tons of feed to produce enough pelts for one mink coat.

Husbandry difficulties on fur farms include fighting among the animals, which can damage the pelts and reduce their value; and poor hygiene and nutrition, which again may impair fur quality. Although

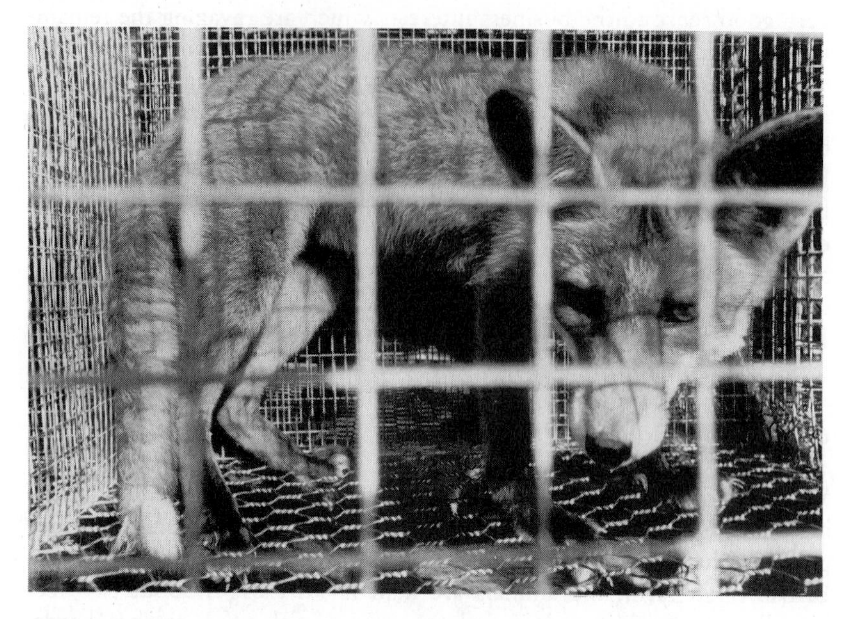

The battery cage system is not only applicable to food animals, but is also used extensively on fur farms. Some 25 million mink and over 2 million foxes are reared each year for the world fur trade. (Mark M. Rissi)

they have been bred in captivity now for up to thirty generations, the reproductive performance of fur animals is not always up to expectations. Infanticide among foxes can be as high as 50 per cent and while the causes are not definitely known, research at the Agricultural University of Norway indicates 'stress' as a probable factor; and provides the interesting information that the cub's tail is usually bitten off first, followed by the hind legs, hind body and fore parts – the exact opposite of the normal procedure during the killing of prey.[3] Work continues to develop a design of cage which, it is hoped, will make the vixens feel less mixed up.

While the breeding of foxes may be a bit of a nightmare, the proverbial fecundity of the rabbit must be any battery farmer's dream. Unlike other livestock, female rabbits can be re-mated immediately after giving birth, and since pregnancy lasts a mere thirty-one days, it is theoretically possible for one litter to be weaned at four weeks old, and another litter to be born just three days later. In practice, does are often allowed three weeks between dropping one litter and conceiving the next, but with an average litter size of eight, this still allows one doe to mother over fifty-two young in the course of a year. Battery farm layout follows the usual pattern of wire cages arranged either in tiers, or more often on the 'flat deck' system, in long, low buildings. If they are being reared for meat, the young rabbits after weaning are sometimes transferred to colony cages, where a stocking density of half a square foot each will discourage them from running about, and ensure maximum weight gain. The aim is to produce a six-pound rabbit in nine weeks, and this requires a concentrated diet of cereals, soya, lucerne meal and meat or bone meal. Some hay may be added as roughage, and the antibiotic virginiamycin can be used to stimulate growth rate. There are also important markets for rabbits in the fur trade, and as experimental animals for the testing of cosmetics and other household products. So far, rabbit farming in Britain is on a relatively small scale, but in France an estimated seven million animals per week are produced, nearly all for meat.[4]

Many other erstwhile game animals are now bred for food using intensive farming methods. Quail are kept in battery cages to lay eggs which may be sold to delicatessens, or hatched in incubators to produce chicks for rearing. After six weeks in indoor pens, the young quail are killed to supply the brisk demand from the restaurant trade for this 'luxury' meat. What they taste like I can only guess; but analogy with the tender but tasteless carcasses of immature pheasants, that have obviously spent most of their lives loitering round the corn bins waiting to be fed, does not inspire hopes of gastronomic delight. Similarly, how much of a treat is venison from captive or indoor herds, compared with the heather-nourished flesh of a genuine monarch of the glen? Such considerations weigh lightly with most British consumers: for many customers the name, rather than the flavour, of unaccustomed luxury foods is their selling point; while most caterers do not care too much about quality once they can feel the price. Only the true sportsman, for whom the finer points of the art still matter deeply, and the poacher, who pays nothing, can afford to be choosy about such niceties of taste.

On a typical deer farm, the animals graze outside during the summer, and are kept under cover in winter, being fed on hay and concentrates containing fish meal, vitamins and minerals. The close proximity in which the animals are kept may encourage transmission of disease, and British deer farmers worried by the introduction of tuberculosis in imported stock have been pressing for a government slaughter and compensation scheme. Some farmers prefer to slaughter their own deer by shooting in the field, but there are commercial pressures for abattoir slaughter, which is more convenient for processing the carcasses, and claimed by some to be more hygienic. Against this, rounding up, de-antlering and transporting deer to the slaughterhouse are difficult operations to carry out humanely, or indeed without serious risk of broken legs and other injuries. Abattoir slaughter caused considerable controversy in Britain recently, as welfare organizations attacked and thwarted the deer industry's proposals to 'regularize' the killing of deer in slaughterhouses.

In the United Kingdom there are around 5000 deer on farms, but in New Zealand the number is in the region of 200,000. Abattoir slaughter is the norm in New Zealand, as is another practice which is banned at present in Britain: the sawing off of the sensitive growing antlers for 'velvet', which is prized in certain countries such as Korea for its supposed aphrodisiac and medicinal qualities. World demand for venison is claimed to be very buoyant, with Germany, Japan and the United States being major importers.[5]

Fish, like game, were traditionally caught in the wild, though the carp kept by medieval monks in ponds convenient for the kitchen, might perhaps be said to have been domesticated. Carp farming was probably practised in China 4000 years ago, and the first treatise on the subject dates from 475BC. The practice was introduced from Europe to England in the fifteenth century. About 100 years ago brown trout began to be reared for the re-stocking of natural fisheries, and the American rainbow trout was introduced as a species for farming. Only comparatively recently has fish farming become a major industry, particularly with salmon and trout, for which production in British fish farms now exceeds the catch from the wild. With both these species, fertile eggs are hatched in controlled environment buildings, on trays which must be provided with a copious supply of fresh running water. When the young fish reach a certain size they are transferred to tanks or to netting 'cages' suspended from rafts moored in a suitable patch of water. Since they cannot forage, they must be supplied with specially formulated feeding stuffs. Trout are usually grown to the required size in fresh water, but salmon are moved to sea cages for two years or more before they are ready for market. A typical salmon cage is around 15 m square; when the fish are fully grown there may be up to six large fish per cubic metre of water.

Salmon farming was pioneered by the Norwegians, who are still the largest producers, but the output of farmed salmon in Scotland has been expanded from just over 1000 tons in 1981 to an estimated 60,000 tons in 1991.[6] This dramatic growth has been largely due to investment by the Highland and Islands Development Board and the European Community's Regional Development Fund, who see fish farming as an ideal way of bringing employment to depressed rural areas. Without such subsidies, the profitability of the industry would be much less certain, because of the high costs of feeding, capital plant and labour, and the lower market value of farmed fish, which are vastly inferior in quality to prime wild salmon. As with furred and feathered game, the flesh of farmed fish differs in texture and flavour from that of the wild animal, due to lack of exercise, unnatural diet, etc.

Production costs are increased by the need to protect fish in cages from predators such as seals and herons, either by elaborate double layers of netting or by shooting the offending creatures; by the need to keep cages free from other marine growths by chemical treatment or mechanical cleaning, and by the usual disease problems of intensive husbandry systems, which are controlled by the use of insecticides, vaccines and antibiotics. However, the ability of large producers to guarantee regular and consistent supplies of fresh or smoked salmon to the supermarket chains makes the product attractive to the large retailers, and barring unforeseen problems with disease or feeding costs, fish farming appears to have a secure future.

It would be wearisome to describe the systems used for factory farming of coypu, crocodiles, chihuahuas or chimpanzees. Suffice it to say that all these species, and many more, can be and are reared using variations of the techniques already outlined, to provide human requirements for fur, skins, flesh, or subjects for vivisection. In the case of such exotic species, their confinement and captive breeding is an abrupt change from the previous norm of direct capture of mature animals from the wild, rather than a more intensive continuation of established farming practice. As remarked already, the new status of these creatures as domestic chattels may seem perturbing, and there is resultant confusion about their welfare requirements. We have been accustomed to regard wild and domestic creatures in a somewhat different light; so how should we react when yesterday's wild animals are today's farm livestock?

The details of how much different animals suffer in captivity, and how we assess this suffering, will be dealt with later. Suffice it to say that from this point of view it is actually easier to decide on a datum for comparison of welfare, or indeed of product quality, if the intensively kept animals belong to a species which is still extant in the wild. Studies of free-living populations can provide information about natural behaviour patterns to which the behaviour of captive animals can be contrasted. For animals with a long history of captive breeding, such comparisons will be less easy, because the domestic species will no longer be genetically equivalent to their wild counterparts, if indeed the latter still exist; while for the 'new' species produced by genetic manipulation, there will be an even more perplexing gap.

From the ethical rather than the scientific standpoint, the parallel existence of similar animals in the wild and in captivity poses some awkward questions, which welfare legislation has not entirely resolved. The opponents of the bill to regularize abattoir slaughter of deer took exception to the fact that, at an arbitrary judicial stroke, the protection to which captive wild deer were entitled under the 1911 Protection of Animals Act was to be replaced by more detailed, but in their view less adequate, codes of conduct, analogous to those already in force for other farm animals. If the status of captive wild animals and domestic ones is really different, as many people feel, how is this to be explained? Status is a social creation: custom, which can be relied on to justify the status quo, can give no account of the way in which domestication seems to change things; nor need we seriously consider the view that divine revelation offers the last word on whether or not the same law is appropriate for the lion and for the ox. Perhaps consideration of an actual example, which illustrates fairly general attitudes, will clarify the issue.

In January 1984 a pet-shop owner in Liphook was fined over £1000 for keeping birds of prey in cramped conditions, which provoked this comment in an animal welfare magazine:

> It was said that an eagle owl did not have enough room in its cage to spread its wings. The cage width was 75% of the bird's wingspan.
>
> Had it been a chicken, of course, Mr Terry could really have packed it in. A 30 inch wingspan chicken can be put into every 4 inches of cage-width.
>
> Then he would not have been fined, he would have been given a subsidy.[7]

Such juxtaposition makes preferential concern for the wild animals appear farfetched and illogical; but nevertheless, the law in this case is to some extent a reflection of general public opinion. If he were tackled about such apparent unfairness, the farmer might point out that his chickens are his property, for which he has paid good money to a breeder, and which only exist at all for the purpose of providing human beings with eggs. Surely he can do as he will with his own?

Although the farmer's conduct is sanctioned by at least the tacit consent of society, his response involves two very common assumptions, neither of which stands up to close scrutiny: that the use we make of animals proves they are 'intended' for human service; and that a person has an absolute right to do as he will with his own property.

It is a common conceit that the natural world was created solely for the benefit of mankind, and if the world was made for man, then he could reasonably do what he liked with it. Such an attitude, albeit somewhat tempered, still represents a perfectly orthodox Christian tradition. But if this is really how we feel, and such an outlook could obviously justify almost any treatment of battery chickens, then what grounds are there to engage in special pleading for eagle-owls? Such special pleading would certainly have been incomprehensible to the medieval mind, but our own age does not place mankind quite so firmly at the centre of creation.

By the eighteenth century, the arguments of sceptics who doubted man's unique right to the bounty of nature were being given further weight by observations of hitherto unknown forms of life and undreamt of celestial bodies through the new microscopes and telescopes, and by the general realization that fossils showed the earth had existed long before the appearance of man. The English Romantic movement took things a stage further, claiming that true spiritual purpose could be found, not in the exploitation of nature, but by approaching it with a sense of reverence and awe:

> . . . a sense sublime
> Of something far more deeply interfused,
> Whose dwelling is the light of setting suns,
> And the round ocean and the living air,
> And the blue sky, and in the mind of man:
> A motion and a spirit, that impels
> All thinking things, all objects of all thought
> And rolls through all things.[8]

The weary town-dweller of today is following the lead of the Romantics as he finds refreshment in a ramble through scenic countryside, or by watching the proud and free lives of wonderful wild creatures portrayed on television. This perception of wild nature as a source of spiritual values is no doubt partly responsible for the different importance we attach to the rare bird of prey and the common domestic fowl.

It is very wholesome, no doubt, to find God in oceans and sunsets, in daffodils and eagle-owls, or in the traditional Christian love of our human neighbours. But if we follow Blake, rather than Wordsworth, we may begin to perceive that it is sheer snobbishness to sigh for caged robins and wounded larks, if we have no feelings for the clumsy dark-reared pigeons that went into the game pie. However, once we start seeing the divine presence in our neighbour's pigs and chickens, there is likely to be trouble.

The purpose we find in other creatures is such disputed territory that it is a poor place to begin an argument. If it is held that the proper purpose of the eagle-owl is to edify and instruct, while that of Farmer Giles's battery chicks is simply to give us a cheap breakfast, these are mere projections of personal opinions. Others may feel, with the eighteenth-century visionary Henry Baker:

> Each hated toad, each crawling worm we see,
> Is needful to the whole as well as he.[9]

Proper purposes can be found everywhere or nowhere; the conservationist finds sermons in stones, where the mining prospector sees only money. To each individual they may seem real enough, but because of their highly subjective nature, such ascribed purposes will play no further part in my arguments.

Perhaps the concept of property will serve better to differentiate between domestic and wild species. Are not the farmer's chickens his legitimately, to dispose of as he will, while the pet-shop man's owl is a wild creature that belongs to no-one? But so, once, were the chickens' forebears, and so were many of the deer and arctic foxes now in captivity. And it is very probable the bird dealer paid for his captive with just as good money as the farmer gave for his hens. Surely if the concept of slavery is accepted, it is illogical to object to the process of enslavement.

Property rights, in any case, are not absolute. Ever since the writer of Deuteronomy forbade muzzling the threshing ox, many societies have set limits to the rights of an animal's keeper to the unimpeded enjoyment of his property. If his idea of enjoyment is throwing cats on to bonfires or kicking piglets to death, we currently prohibit such activities in ways we should not apply to a person who merely burnt a piece of furniture or smashed crockery. Such restrictions apply both to keepers of wild animals, and to pet owners and farmers, though not to all with the same severity. It is hopelessly out of date to object to the concept of 'animal rights', when existing welfare legislation already grants such rights, however limited or conditional. The trend today is very much towards a progressive extension of such protective legislation, and to a more general recognition of the right of all animals not to be treated cruelly for malicious or trivial reasons. As with human rights, which cannot practically be unlimited, it is up to society to decide how far it can or will go – though some people would claim that legislation in favour of animals is unfairly hindered by their lack of voices or votes. What does seem clear, despite accidental inconsistencies of legislation, is that we are prepared to be more generous in this respect to some species than to others: equality under the law is not part of the animals' deal.

Regrettably, it must be conceded that our common practices are based less on principles than on expedience. We have at least re-learnt a certain reverence for the natural world which our urban and exploitative cultural tradition had almost lost, and most of us get a greater thrill from eagle-owls soaring free in the wild than any pleasure we might have from them behind bars. This is a very recent phenomenon, which has been greatly encouraged by the way television takes us into the wilderness, rather than just collecting little bits of it in the way of zoos and pet shops. But while such changes of attitude are welcome, how much harder to extend a similar warmth of feeling to battery fowls, which are so useful where they are. We are more generous in the rights we allow wild animals, particularly rare ones, not on grounds of logic, nor even on grounds of custom (witness the rapid domestication of deer or salmon), but quite simply because wild creatures are of less general use than their domestic counterparts.

Property rights, deriving from the general will of society and circumscribed by the same, deserve none of the superstitious reverence they sometimes gather, and cannot be regarded as fundamental issues in the animal welfare debate. In a democracy, there is no particular reason why the rights of owners or keepers of livestock should not be restricted in a way consonant with the general sentiment as to the desirable degree of protection against cruel or negligent usage. There is little evidence that such sentiment differentiates between the rights of owners who have bred their own livestock, and those whose title comes by purchase, gift or capture. It is tempered, rather, by the perception that oppression of some species has a greater utility to society at large. Nevertheless, it is tempting for the animal breeder, or still more the genetic engineer, to claim that his creative involvement in the breeding process gives him a greater right to use the resulting beasts as he will. 'New' species may not always be covered by existing legislation, and there will inevitably be efforts to exploit any available loopholes, as appears to be happening in the United States, where genetically engineered laboratory animals are not covered by existing welfare regulations. There is, however, no clear precedent for pandering to special pleading of this sort. I am not aware of any society which has differentiated in this way between the rights of a breeder and a mere owner of livestock. We do, of course, claim additional rights over human beings we have ourselves begotten; but surely not on grounds of the creative effort involved! The analogy is not precise: the proprietor of animals is more akin to a slave owner than a parent, and as such is burdened with some rather minimal duties towards his charges which are quite independent of whether or not he superintended their conception. Claims about proprietorship, of individual animals or of patented procedures for genetic manipulation, are best kept separate from the discussion of the rights of animals to treatment without unnecessary cruelty, which should apply equally against owners, breeders, genetic engineers and mere passers-by.

The nature of animal suffering will be discussed in more detail later on, as will the need for various practices which cause such suffering, measured in terms of utility to human society. These concepts, it is hoped, will be more useful than the contentious idea of preordained purpose, or the peripheral notion of property rights, both of which merely confuse the main points at issue.

Man's dominion over food animals is actually exercised by farmers, who are only a minority of the population. While the rest of us may play a democratic part in defining and protecting the rights of their livestock, for many people the main contact with farm animals is in the butcher's shop or on the dinner plate. In this case, animal welfare is far from being the most obvious concern which comes to mind. We all have selfish but quite legitimate preferences for certain types of food, on grounds of social or religious custom, flavour or ideas about what foods promote good health. Even from these basic consumer angles, the products of intensive livestock systems seldom compare favourably with those from traditional farms.

Part II
PRACTICAL CHOICES

4

The Choice of Diet

The apostle Peter had a vision in which a great sheet was let down to earth 'wherein were all manner of four-footed beasts of the earth, and wild beasts, and creeping things, and fowls of the air'. In his vision, we are told, God said to Peter that he should no longer consider the creeping things, etc., as unclean, and the followers of Christ were thus released from the Jewish laws about clean and unclean meats. For those who are interested, the eleventh chapter of Leviticus gives a comprehensive run-down on which foods are tabooed: locusts, beetles and grasshoppers are permitted, but the lapwing, the bat and 'all fowls that creep, going on all four' shall be an abomination to the righteous. So shall rabbits, hares, pigs and 'every beast which divideth the hoof, and is not cloven-footed nor cheweth the cud'.

What is the meaning of this classification of food as clean or unclean? Although anthropologists report food taboos in virtually all cultures, the explanations they offer vary from reasons of health or economy to capricious superstition or the definition of exogamous tribal groups. Whatever their underlying justification, the seriousness with which such dietary restrictions are regarded in virtually all cultures can be difficult for the modern liberal outlook to comprehend. We can understand the avoidance of foods which carry a risk of poisoning, but the arbitrary nature of many taboos makes them appear absurd in the cool light of reason. Not that this stops us observing our own dietary customs; few British people would enjoy a meal of dog, regardless of any rational justification they might give for their repugnance. But the hold of tradition is slackening; in an increasingly cosmopolitan world British holidaymakers tuck in to snails and frogs' legs, and diners in Israel are offered pig meat thinly disguised as 'white lamb' to save their feelings.

What we eat is limited by availability as well as by religious scruples, cultural traditions or individual squeamishness. In the affluent West, the choice has never been wider, with supermarkets crammed with goods from all over the world, and a range of ethnic restaurants in most cities.

But do we use this choice wisely? Many nutritionists would argue that we do not, and that there is an ever-increasing weight of evidence connecting twentieth-century ailments such as coronary heart disease and cancer with dietary factors. Even with an ample range of foodstuffs in the shops, the public's choice of diet may be unhealthy for a variety of reasons. These include misguided ideas about nutrition, addiction to particular foodstuffs, limited housekeeping funds, the attractions of convenience foods, and the deliberate efforts of advertisers and manufacturers to mislead the consumer.

A knowledge of nutritional theory is by no means essential to survival. In most times and places people have learnt what foods are suitable by the same methods as are used by other animals: trial and error, and instruction by their elders. Experiments in which young infants were presented with a wide choice of foods without parental guidance have shown that left to themselves they can do surprisingly well at selecting an appropriate diet. As we grow older, the human love of theorizing may lead us further astray; as 'knowledge' derived from overhasty generalization turns out to be inadequate or misleading.

For centuries, fruit was considered unsuitable for children, as it was supposed to cause fevers and 'fluxes', and when the tomato first arrived in Europe it was widely suspected of causing cancer – as well as being a notorious aphrodisiac. Such quaint ideas make us smile, but are we sure that today's fashions in diet are more rational? In 1950, the conventional wisdom of the day was that 'those who can afford a good variety, and generous supply of first-class protein, will have little need to consider rules of diet'.[1] The same author expressed grave doubts as to whether a vegetarian diet 'is adequate for full mental activity, physical vigour and resistance to disease'. More recently, it has become clear that scientific evidence overwhelmingly refutes such claims, which are scarcely defended now by anyone outside the meat trade.

But old habits die hard, and the association of meat foods with feasting and plenty is deeply ingrained in our culture. Meat used to be a seasonal luxury before the Dutch innovation of storing turnips for winter feed made it economic to keep more non-breeding animals alive during the winter months, so the association is reasonable, but like other luxuries, there is a limit to how much is good for you. When only the wealthy classes tucked in regularly to large flesh meals, their attacks of gout and apoplexies were seldom blamed on any specific ingredient of an overindulgent life style. Several recent studies, however, have produced evidence that a diet rich in animal fats carries substantial health risks.

Like the link between smoking and lung cancer, the connection between animal fats and heart disease is circumstantial, and based on the interpretation of complex statistics, so it would be foolish to present a small batch of plausible results as 'proof' of the dangers of a high meat diet. However, as the evidence has accumulated, a number of independent reports have emphasized the high probability of animal products causing heart disease. The American Heart Association has recommended cutting down on animal protein and fats (including dairy produce) and eating more fruit and vegetables.[2] In Britain, the National Advisory Council on Nutrition Education recommended a cut of about a half in saturated fat intake, as well as a reduction in salt and sugar, and increased dietary fibre; while the Committee on Medical Aspects of Food Policy advised a 25 per cent cut in saturated fat consumption.[3]

Certainly, meat consumption in the rich Western nations has reached an all-time high, though there is now a trend towards eating less red meat and more poultry. In the United States, beef consumption in 1972 was 53 kg (116 lb) per person per year, exactly twice the 1940 level; but by 1985 people were eating around 50 kg (108 lb) of beef, and a record 34 kg (75 lb) of poultry meat.[4] The comparable figures for the United Kingdom are about 18 kg (40 lb) of beef and almost the same weight of poultry, and show a similar trend towards white meat at the expense of red.[5] It remains to be seen whether eating broiler chickens and farmed fish will actually be better for the nation's health than the traditional roast meat and two veg.

Heart diseases are not the only ailments which have been linked to a high meat diet. Causal connections have been proposed with breast and colon cancer, and there have been a number of studies which suggest a better general level of health among vegetarians than among meat eaters. Useful data has been obtained from religious groups such as the Seventh Day Adventists, who eschew meat, tobacco and alcohol, and the Mormons, who avoid smoking and drinking but do eat meat.

Immoderate meat eating is just one of the elements of the modern Western diet which has been criticized by doctors and nutritionists. Excessive use of salt is claimed to aggravate high blood pressure, while milk is a common cause of allergic reactions such as asthma, insomnia and depression. Refined sugar and white flour have been implicated in a variety of health disorders including dental caries, diabetes and heart disease.[6] Arguments will doubtless continue over what constitutes the ideal diet, between those who regard fat, sugar, salt and refined flour as the four 'white poisons', and their rivals who claim that such charges are unfounded, or at least unproven. But even the staunchest defenders of these products concede that while a little of what you fancy may be good for you, it does not necessarily follow that more will be better.

The facts are that consumption of 'unhealthy' foods has peaked in the West, though it expands apace in many of the developing countries. Consumers in the affluent nations are beginning to appreciate the good points of the traditional peasant diet of rough bread, a lot of vegetables and a little meat, while in societies which are just entering the sliced bread stage of evolution, such fare is despised as a reminder of poverty and hunger.

An increasing number of people are now rejecting meat and fish altogether, and adopting a vegetarian diet. Their reasons vary from the purely selfish motive of safeguarding personal health to a variety of altruistic and spiritual aims. At present, an estimated 1.5 million Britons are vegetarians, of whom about 60 per cent are women.[7] There is no evidence that a properly balanced vegetarian diet (with or without dairy products and eggs) is in the least inimical to good health, and numerous vegetarian cookbooks are available for those who wish to be initiated. With the variety of produce now on sale in most supermarkets, vegetarian food can be made very interesting and digestible, though most meat eaters might find an abrupt transition less easy to cope with than a more gentle change by gradually eating less and less meat.

For many people, the most important bodily effect of what they eat is the impact on their physical appearance. Fashions in slimming diets are at least as volatile as ideas about food and health, with starch 'out' one year and 'in' the next. Not surprisingly, few diets recommend fatty food, and the quest for beauty may be as much responsible for consumer resistance to fatty meat as the quest for health. Fat content depends very much on the method of presentation: roast chicken flesh has a fat content of 5 per cent, compared with 7 per cent for roast pork, 9 per cent for roast beef, and 11 per cent for roast lamb; but roast chicken with the skin left on contains at least 13 per cent fat,[8] and many prepared foods such as sausages and hamburgers contain just as much as the manufacturers believe they can get away with. Any correspondence between the demands of vanity and health cannot be pushed too far: many slimmers suffer from ulcers, and about one in 600 schoolgirls develop anorexia nervosa.

Fashions change. Although affluent Westerners now eschew fatty meat, for centuries domestic animals have been bred to give fuller, fatter carcasses. Thus a modern British pig has a carcass fat content of around 40 per cent, compared with just over 1 per cent for a free living warthog; while lean beef cows have 25 per cent fat as against 2 per cent for the nomadic cattle of East Africa.[9] In recent years, breeders of meat animals have made great efforts to produce animals which still grow rapidly, but produce more lean meat and less fat, but there are limits to how far they can go if the other profitable features of modern stock, like rapid growth rate, are to be preserved. Very lean pigs, for example, bruise easily and are susceptible to stress. Worse from the farmer's point of view, they have a greater tendency to produce meat of poor texture, unpleasantly described in the trade as 'pale, soft and exudative'. Research is now being directed at an alternative to selective breeding: the injection of chemicals which prevent fats being laid down in the normal way. Lovers of pork crackling, look out!

Conscious dietary aims exercise a negative influence on the foods people choose, as they try to avoid unhealthy or fattening substances; but while food producers try to react to these aims, or at least play down the dietary disadvantages of their own products, there is also intense competition between manufacturers to provide food that people positively like. Tastiness, visual attractiveness and convenience are examples of the positive attributes that processors and packagers want to emphasize, and they do this in many ingenious ways. As anyone who has tried to cut the sweet eating habit will know, sugar is mildly addictive, and manufacturers exploit the sweet tooth of the public by lacing an extraordinary variety of processed foods with large amounts of unnecessary sugar. For those hooked on salt, salty foodstuffs will have a similar attraction, while at a slightly more subtle level 'flavour enhancers' like monosodium glutamate can be equally effective. Lobbying by consumer groups has at least resulted in labelling regulations whereby the more literate public can get some indication of what is in the tomato ketchup or the stock cube, and the enormous use of additives in processed food has been exposed by the European Community (EC) E numbers which now appear on the packages.

Many food additives have been incorporated to increase the appeal of food to shoppers, and some manufacturers seem rather bemused by the change in attitude of the fickle consumer, now the sinister looking E numbers appear on the packet. We can spare them too much sympathy, though, since many more additives merely simplify processing or prolong shelf life, and in view of their possible dangers it is surely fair that people should be able to find out what they are buying. And how much does an additive, designed to make food more attractive than it naturally is, help the consumer? Before giving too much credence to the bland public relations of the food processors, or to the disingenuous labelling and advertising of their products, we should remember that their industry has a long and notorious history of dishonest practice.

As long as food has been bought and sold, there has been an incentive for the merchant to deceive his customer over the quality of his goods.

In Ancient Rome, Pliny reports the adulteration of bread with 'white earth', and other classical writers refer to inspectors who had the duty of supervising the quality of wine or meat. After the Norman Conquest of Britain, a special law was passed forbidding the adulteration of bread, but the frequent literary references to dishonest millers make it clear that the problem did not go away. In the eighteenth century, white flour was often sold with the addition of chalk, lime or even white lead, while foreign and often poisonous substances were regularly incorporated with tea, coffee, sugar and milk, as well as in made-up products such as pies and sweets. In 1820 Frederick Accum published *A Treatise on Adulteration of Food and Culinary Poisons*, in which he gave the names of convicted merchants with an analysis of their wares, but it was not until 1875 that the Public Health Act obliged local authorities to appoint public analysts and to carry out regular sampling of foodstuffs.

The 1875 Act, and its successors up to and including the 1984 Food Act, do not provide a comprehensive list of substances prohibited in food. Preservatives and colouring agents are among the additives which are currently controlled by specific regulations, while most food ingredients and flavourings are not restricted; except by the general duties of the manufacturer and seller not to include harmful substances, and not to use deliberately misleading descriptions. Since the issue of additives was brought into the public eye by the 1984 EC labelling regulations, there has been a substantial consumer revolt against their proliferation, to the extent that 60 per cent of respondents to a 1986 survey had no faith in official controls, and 80 per cent felt that many additives in use were harmful and should be banned.[10] This sudden awakening of public consciousness came after a period of intense development of the food processing industry, during which the quantity of additives used in Britain increased from about 20,000 tons per year in 1955 to 200,000 tons in 1985.[11]

The manufacturers' reasons for this large scale use of additives need not be imagined to be philanthropic. For the estimated £300 million they spend annually on added ingredients, they expect a good return, while benefits to the consumer are nugatory. Of course, extending shelf-life may keep prices down, and there would be dangers in removing preservatives from sausages without massive re-education of the public, but preservatives account for less than 2 per cent of all additives consumed in the United Kingdom.[12] The remainder, with the exception of a relatively small number of 'processing aids' which stop foodstuffs sticking to machinery, or in some other way facilitate manufacturing processes, have functions which are entirely cosmetic. These include colours, flavours and texture modifiers, all of which make processed food more 'attractive' to the shopper, though the survey results quoted suggest that the attraction is somewhat diminished if the punter in the supermarket is aware of what he's buying. (Or, just to maintain conventional sex stereotypes, if the housewife is aware of what she's buying: the Presto survey was, in fact, aimed at female shoppers.)

But whatever the masculine or feminine consumer makes of them, additives are the food industry's delight. What could be more attractive from the profit angle than the extra ingredients so thoughtfully added to the average banger? A typical supermarket sausage might contain E223, E450, E301, MSG(621) and Red 2G(128), as well as unspecified flavourings, of which several thousand are at present completely uncontrolled. Decoded, E223 is sodium metabisulphite, a long-established preservative which discourages the growth of food poisoning organisms and the development of rancidity. Although sulphites are very widely used, they have been linked with asthma attacks, and with mutations in bacteria. They also destroy vitamin B1 and have a bleaching effect, which justifies the use of artificial colouring. E450 is a polyphosphate emulsifier, beloved by the manufacturers of sausages because it binds extra water into the meat fibres; increasing the weight of the product at negligible cost, and zero benefit to the consumer. E301, sodium L-ascorbate is an anti-oxidant, which stops fat turning rancid; it also discourages colour fading. Monosodium glutamate (621) stimulates the taste buds, giving an exaggerated impression of flavour. It is banned in baby and infant foods, as it has been linked with changes in brain chemistry, and is known to cause dizziness and palpitations in some subjects. The red dye which completes the list is used to colour the fat and rusk components of the sausage so that they look like meat. Red 2G is used in preference to other permitted colours because it does not fade when exposed to the bright lights of display cabinets, but it has been accused of causing genetic damage which can lead to birth defects, and has been banned in the United States, Austria, Belgium, Denmark, France, Norway, Sweden, Switzerland, West Germany, Japan and Canada.[13]

Where, one might wonder, is the benefit to the consumer from all these excellent substances? It is very much in the producer's interest to pass off an inferior product as fresher, more substantial, tastier and meatier than it actually is; but while this is quite legal, it must be open to grave doubt whether it is decent or honest. The Vikings did not distinguish between unfair trading and theft, but as Clough puts it, we are rather more sophisticated:

> Thou shalt not steal, an empty feat
> When it's so profitable to cheat.[14]

Of course, when cheating is legal and respectable, it can only be expected that legal and respectable corporations will indulge in it and call it something else. And worse, those which do not are likely to lose out to the others in terms of sales and profits.

Consumer resistance has had significant impact on the use of additives over the past few years, and pressure is being exerted by such groups as the Food Additives Campaign Team for more comprehensive labelling, and the banning of more substances of dubious safety. However, despite such reforms, unless the food manufacturing industry as a whole starts valuing real quality above the mere appearance of quality – and the evidence of the past 2000 years or so is not too hopeful – we must expect a constant stream of stratagems for evading protective legislation.

Whatever justification can be made for the numerous food additives in common use, in terms of making products keep better or appear more appetizing, they cannot possibly apply to the oldest and simplest of all types of adulteration: the addition of water. Illegal watering of beer and milk still account for a large proportion of prosecutions against dishonest traders, but the quite legal incorporation of excess water in fish and meat products represents an even more widespread fraud against the consumer. The problem is not easy to come to grips with, because the majority of foods naturally contain substantial amounts of water: 40 per cent in bread, 75 per cent in raw lean beef, 82 per cent in raw cod fillets, and up to 96 per cent in a fresh lettuce. The natural water content is not, of course, fixed at an exact percentage, but may vary quite widely according to produce quality or season. This makes dishonest addition of water difficult to detect, and difficult to legislate against in terms of a maximum allowable water content.

Limits have been introduced to check the flagrant addition of water to frozen poultry. Oven-ready chickens are cooled in a bath of iced water before being frozen, and if they are not given reasonable time to drain, anything up to 40 per cent of the weight when sold may simply be frozen water. European Community regulations now stipulate a maximum 7.4 per cent added water for chickens, but do not apply to turkeys, ducks or geese, for which a 1986 survey found up to 19 per cent water retention.[15] Fish products are also vulnerable to this type of adulteration: a thin layer of 'ice glaze' protects frozen fish from deterioration in storage, but while a maximum of 10 per cent added water will achieve the desired effect, some manufacturers regularly add 50 per cent or more water to frozen prawns, making fair price comparisons virtually impossible.[16]

Many fish and meat products have extra water incorporated by treating them with polyphosphate solutions, E450. These chemicals are used to stick together the small pieces of fish and meat used in re-formed products like fish fingers and turkey steaks, where their effect on protein solubility is a significant contribution to the process. But they are also widely used to make meat 'succulent and tender'; to 'improve carving characteristics', reduce 'unsightly drip' on thawing, and to cut down 'cooking losses'. Translated into plainer words, these phrases all amount to more water and less meat. Most cured meat products are injected with a pickling solution containing polyphosphates; so that while a traditionally-made ham loses weight during curing and again when cooked, the typical supermarket roast ham contains up to 20 per cent more water than the raw meat from which it was made. The process of reshaping meat from pieces which have been tumbled and massaged in phosphate solution before being formed into ersatz joints is a particularly effective way of incorporating extra water. Declaration of added water is now required in the small print of the label, but there is no limit to how much is permitted.

'Cooking losses', which do not concern the home cook unduly, since all that is lost is water, are of great interest to the purveyor of ready cooked meat. A roast chicken joint prominently labelled 'no added water' may turn out to contain the ubiquitous E450, so that while water has not actually been added, the end product is certainly wetter (and heavier) than if it had been untreated. And if it has been 'flash roasted' rather than cooked in the traditional hot dry oven, cooking losses are reduced even further. 'Flash roasted' meat has actually been cooked by steaming gently, followed by very brief browning of the outside in a high temperature oven, which does not cook but merely colours.

It must have become apparent, that even with the clearest idea of what he wants to eat, the consumer faces a difficult task comparing the offerings of the food processing giants with any sort of objective standard. It may be felt that some of the more patently undesirable forms of adulteration commonly practised should be banned, or one can argue that the buyer should beware, provided that labelling is comprehensive and reliable. While some of the food barons are prepared to use every trick in the book in the pursuit of profit, and are frankly rogues, others are less disingenuous about their wares, and make genuine efforts to offer the public the chance to make an informed choice. Many of the supermarkets are commendably on the side of the consumer, and have taken a constructive part in promoting better labelling and better choice. Perhaps the honest merchants can take some comfort from these quaint lines taken from *The Experienced Butcher* published in 1816, and quoted by Gerrard (1977):

> When a tradesman is known for upright dealing, he will have all the custom worth having, and his gains, if not so great as some, yet it will be the more pleasant in the getting, more satisfactory in the using, and alone delightful in the thinking on.[17]

For the majority of people, the risk of being cheated is less alarming than the risk of being poisoned, and the recent increase in public awareness concerning additives and residues has been almost entirely due to concern about their safety. However, evaluation of which substances are safe to eat is extremely difficult. In 1902 Dr Harvey Wiley conducted tests on a number of additives for the US Department of Agriculture by the attractively direct technique of feeding the chemicals in question to a 'poison squad' of volunteers. This work has been quoted by Bernarde (1975):

> I wanted young, robust fellows with maximum resistance to the deleterious effects of adulterated foods ... if they should show signs of injury after they were fed such substances for a period of time, the deduction would naturally follow that children and older persons more susceptible than they would be greater sufferers from similar causes.[18]

While Dr Wiley's research did result in the banning of certain food additives, it must be admitted that such crude methods have their limitations, particularly for evaluating long-term risks.

Obviously, some chemicals are so immediately poisonous in small doses that it would be rash to allow their use as food additives, but many common and even essential items of our diet are toxic in large doses, though safe enough at normal levels of consumption. For example, arsenic is present in many foods in small quantities without posing a significant health risk. Chronic toxicity, of small doses fed over a long period of time, may be difficult to establish for certain; as it is normally impossible to find two otherwise similar groups of people, only one of which has been exposed to the particular substance in question. Long-term effects of carcinogens, which cause or promote cancer, can show up long after exposure to the toxin has ceased. Other long-term effects are also significant: teratogens cause reproductive problems including birth defects of a non-hereditary type, while mutagens damage the genetic material which is passed on to subsequent generations.

In the absence of clear proof, many scientists counsel caution in tions, much of the information available concerning toxicity is based on animal testing, in which rats or mice are fed large amounts of the test substance to establish the normally fatal dose; and small amounts over a long period of time to establish whether there are significant signs of cancer or reproductive abnormalities. The results of such tests correlate up to a point with human experience, but cannot be regarded as definitive.

Evidence from epidemiological studies of cancer must also be regarded as circumstantial rather than conclusive. In only a very few cases, such as the effect of smoking as a cause of lung cancer, have connections been established beyond all reasonable doubt. Currently, debate is raging as to whether or not there is a link between nitrate intake and stomach cancer. Comparisons between different countries suggest that this is the case, but recent studies within Britain have failed to correlate cancer risk with exposure to nitrates.[19] Since over 10,000 people per year die of stomach cancer within the United Kingdom, and there are large variations in nitrate intake from area to area, this gives some indication of the difficulty of pinning down the particular causes of cancers.

In the absence of clear proof, many scientists counsel caution in departing from 'biological normality' without good reason, and similar sentiments are evident among many purchasers of 'organic' and 'natural' foods. But just what is normal, and how safe is it? It would be foolish to take too rosy a view of the wholesomeness of all the foods eaten by our ancestors. Modern hygienic practice and scientific knowledge have more or less abolished several major poisoning risks. Ergot infestation of rye produced spectacular outbreaks of poisoning in the Middle Ages (St Anthony's Fire); and mouldy grains and nuts can still present a serious health hazard. Although the last recorded outbreak of ergotism involved only a handful of people in France in the 1950s, poor storage conditions for these products can allow other moulds to flourish, with the possible production of highly carcinogenic 'aflatoxins'. Before pasteurization, infected milk was frequently responsible for the transmission of serious diseases such as scarlet fever and tuberculosis. In 1845, the milk in London was so bad that microscopic examination invariably revealed contamination by blood or pus.[20]

There are limits to the amount of experience accessible to us, and we can never be sure that some widespread and apparently safe habit is not stealthily doing us harm. The Romans probably never suspected they were suffering from chronic lead poisoning, as has been proposed on the basis of modern archaeological research. And people accustomed to eating heavily smoked or salted foods are unlikely to accept that habits which have been continued over many generations could possibly be dangerous.

There is a presumption in legislation that 'natural' foods are harmless, while imitation or synthetic foodstuffs must be proved safe before they can be used. The American Food, Drugs and Cosmetic Act lays down in detail just which substances may be regarded as 'natural' (e.g. those purified by distillation), and which as 'synthetic' (e.g. those obtained by crystallization). Such hair-splitting definitions can result in discrimination against refined products such as safrole, which is banned as a carcinogen, while its natural source the sasafras root continues to be recommended by the *Whole Earth Almanac* as a beverage and flavouring agent. The food industry is quick to point out such anomalies, and it is a commonplace that many everyday foods would not survive the rigours of the Delaney amendment prohibiting the use of carcinogenic substances in food, or of the 100-fold margin of safety applied to other additives of known toxicity. Thus potatoes might contain solanine, a potentially dangerous nerve poison, with only a ten-fold safety factor; lima beans contain glycosides which break down on cooking or during digestion to produce hydrogen cyanide, and have been responsible for several serious poisoning incidents; and at 100 times normal consumption levels turnips and cauliflower would damage the thyroid, while vitamins A, D and K would all be intolerable at similar levels.[21]

Exact estimates of toxicity are so difficult, that in a quantitative estimate of the validity of animal bioassays for human carcinogens published in 1983 (cited by Abraham and Millstone, 1989) Salsburg was forced to conclude that such tests are more often wrong than right in detecting the toxicity of chemicals known by epidemiology to be carcinogenic in human beings.[22] However, it remains a reasonable policy not to multiply risks by eating large numbers of unfamiliar substances without good reason, and while many additives are doubtless quite innocuous there is unfortunately no way of proving in advance which these are. Similar arguments apply to pesticide residues, which are regarded by many people as a health hazard; in the United States the firm of H. J. Heinz has begun to insist on crops grown without suspect pesticides for its baby food preparations.

Though long-term risks of cancer or other serious illnesses are uppermost in the majority of people's minds, for a relatively small number of people certain food additives cause allergic reactions which, if seldom fatal, can be extremely unpleasant. An estimated three to fifteen people in every 10,000 suffer from asthma, migraine, eczema or hyperactivity induced by common food additives such as azo dyes. Some countries have banned or severely restricted these additives in an attempt to protect this significant minority, but others, such as Britain, take the view that if foods are adequately labelled the individual can look after himself.

But while colourings and certain other additives must now be declared on food labels, many other potentially harmful substances need not be shown. Antibiotics fed to pigs or chickens may remain in the meat when it is sold, as may hormones which have been used as growth stimulants; and the residues from fertilizers and pesticides may contaminate fruit, vegetables or animal produce. Bacterial contamination is another serious problem, particularly with meat, but also to a lesser extent with eggs; and in processed foods there are often residues of solvents and other chemicals used during manufacture. None of these foreign elements need be declared (nor are they likely to be, if present) but public concern about food quality has begun to encourage the development of a retail market in more 'natural' foods, which (by implication or by guarantee) are free from most of the unwelcome extras. At the eleventh hour, a small but influential section of the public is turning its back on the produce of intensive agriculture and high technology food processing, in the hope that traditionally-grown crops and animal products will give better health, as well as more enjoyment.

5
Free Range or Factory?

Most of the food consumed in Britain is 'processed' in some way: milled, cooked, refined, pasteurized, canned, frozen or otherwise altered to give a more conveniently marketable product. Food of doubtful quality, or plain unfit for human consumption, can be treated to render it acceptable (as for example the use of incubator reject eggs in egg-based foods for catering).[1] Only a minority of manufacturers go as far as this; the successful names in the processing industry have achieved their pre-eminence by nurturing public confidence in their products, many of them offering guarantees of immediate replacement if the customer is dissatisfied. Consistent quality is an essential feature in sustaining this confidence, and the food industry has been remarkably successful in standardizing the taste of baked beans, sausages, bacon, etc., despite the inevitable variations in quality of the raw materials. Over 3000 different flavouring substances are currently added to foods on sale in the United Kingdom, and whether the processer wants to make turkey taste of ham, or potato crisps taste of cheese and onion, it can easily be done. Even if he just wants to make frozen peas taste like frozen peas, or tinned strawberries taste like tinned strawberries, technology will lend a hand: if the growers look after size and shape, the manufacturer can deal with colour, texture and taste.

Unfortunately, it is one thing making things all taste the same, but quite another making them actually taste right. In our restaurant at Scarista House we served a no-choice dinner, and one often sensed an air of disappointment when the offering for the evening was plain baked ham. Without fail, however, there was general delight when instead of the flaccid, sugary, chemical-impregnated and practically tasteless substance which often passes for ham, the meat turned out to be firm yet tender, with a complex and delicious flavour such as many had not tasted for years, and others had never tasted at all. The difference between our supplier Ann Petch's Devon hams and their distant supermarket cousins is immense: she keeps traditional breeds of pig renowned for their flavour rather than modern hybrids selected only for efficiency of feed conversion; she rears them outside, without growth promoters and on a wholesome mixed diet; she takes pains to reduce stress at slaughter, which can taint the meat, and she cures the hams by old-fashioned methods and without using the polyphosphates referred to in the previous chapter. All these factors mean a substantial price differential, but the result is a superb product, for which demand continues to expand.

The near demise of old style ham, free range eggs or real ale, and their current revival as luxuries poses some interesting questions. Some people, at least, are prepared to pay quite a lot for food that tastes right, but how do we decide what is the 'right' taste? And are the rest of the public really unable to tell the difference, or are their taste buds simply being overruled on grounds of price or convenience?

The human sense of taste, though very sensitive, is also very easily confused (try tasting a piece of orange before and after sucking a sugar lump, or a dry red wine before and after a sweet white one). The reference level by which tastes are judged depends on what has been eaten or drunk previously, as well as varying with changes in saliva composition according to the time of day or state of health. Comparisons of taste can be made more precise if the mouth is always rinsed with clean water before tasting, but even then we lack a well-developed vocabulary to communicate the experience. The task of the wine or tea taster is made easier by the fact that reference samples can be maintained for reasonably long periods: unless something untoward has happened, a bottle of 1978 Château Lafite tasted now should be pretty much the same as another bottle from the same vintage assayed a month or two ago.

Comparison of foods by flavour has not developed into such a precise study as wine tasting, partly because most foodstuffs are not eaten as purchased, and so much therefore depends on the skill of the cook. An expert chef has a highly trained memory for nuances of flavour, plus the technical knowledge of how to adjust the taste of his creations in the desired way, and he will be sensitive to small differences in the quality of his raw materials. For the less expert of us, however, it is likely to remain a bit of a mystery why this week's *boeuf à la mode* does not seem to taste the same. Was it the raw ingredients, did we boob somewhere in the cooking, or is it simply a trick of imperfect memory? One thing is certain: there is very little point trying to preserve a portion of today's dinner in the fridge or even the freezer as a reference for next time. The taste of freshly cooked food is so volatile that it changes after being on the plate for a matter of minutes. The freshly carved slice of roast beef is very different from the piece which has been kept warm for five minutes, and still more different from the reheated-in-gravy offering characteristic of institutional catering. The evanescent quality of so many flavours we encounter, and the variety of factors responsible for the taste of food on the plate leave many of us rather vague and inarticulate about what we like and why, and lack of practice or interest may extend this indifference even to foodstuffs and beverages which could in principle be compared quite easily.

Lack of practice at taste discrimination may also mean we are more suggestible about flavours than over visual stimuli. Psychologists investigating the sense of taste and smell report problems of this type – even to the extent of people claiming to experience smells which they were told had been transmitted by television. Appearance is extremely important, as shown by experiments in which subjects are repelled by food which is coloured blue by a harmless dye. Other factors which almost certainly influence the perception of taste are the degree of hunger; whether the taster is tense or relaxed, fit or unwell; distraction by background odours; how much the food cost; how it is described on menu or packet; how fashionable it is, and the way it is advertised. When we consider how visual illusions are only observed and resolved through discussion and measurement, and the disputes that arise over the correct naming of colours, it is to be expected that illusions and uncertainties of taste, for which the degree of shared experience and the vocabulary available for description are both much more limited, will be at least as prevalent even if less widely recognized. As with colour perception, there are established physiological differences in taste between individuals; a well known case being the inability of some people to taste phenylthiourea.

Despite all these difficulties, there is a fair amount of common ground about what tastes good. Anyone with only a modest experience of wine soon learns to favour particular regions and styles, though approaches vary from the adventurous to the ultra-cautious. And at the 'quality' end of the market, even allowing for the exaggeration of price differentials by fame and fashion, there is a measure of agreement about the appeal, complexity and balance which justify the high prices paid for top growths. Because of this consensus, much interesting scientific work has been carried out with the aim of establishing what factors of cultivation and vinification influence the final product. Such research has greatly helped the wine makers of California and Australia, who produce some excellent wines; but however much care they take over the choice of vines and the methods of fermenting and bottling, they cannot replicate precisely the climate and precise geology of the great vineyards of Burgundy and Bordeaux.

How different, particularly in Britain, are attitudes towards food quality! Wine producers throughout the world are at least conscious of the importance of quality, which the high value of their product, its portability, stability in bottle, and the large numbers of competitive 'blind' tastings have combined to maintain and improve in recent years. Governments everywhere encourage this concern, with restrictions on methods of production and tight labelling regulations to protect standards. With food production such controls are quite minimal, and entirely negative. There are, it is true, regulations about how much water can be added to milk, and about maximum allowable residues of a number of poisonous substances, but very little encouragement for farmers to pursue quality rather than quantity, and very little protection for those who do have a superior product to market.

The quality butcher, who would choose his own prize beasts on the hoof, the specialist poulterer, and the traditional fishmonger are all endangered species. Volume retailing, rather like the processing industry, tends to favour consistency rather than individuality of flavour. The beautiful Colston Bassett Stiltons we served at Scarista are *never* the same, yet they are invariably delicious; unlike their supermarket rivals which always taste identically of next to nothing. Likewise, wild deer and wild salmon vary, and so do other fish from season to season, potatoes from peat or loam, and so on.

Why, then, is so little attention paid to the differences in taste between wild and domesticated animals, or between intensively farmed and traditional meat, poultry and eggs? Why do food pundits spout such nonsense as the often repeated claim that there is nothing to choose between fresh wild salmon and the farmed product? When a difference of a hundred metres may be all that separates a first growth claret from a *crû bourgeois* (the difference, perhaps between geological formations many feet below the surface) it is impossible that there should be no difference in taste between an active, free living creature and one kept in close captivity on a totally unnatural diet. Yet – with a few notable exceptions such as Derek Cooper and Colin Spencer – this obvious fact passes entirely unnoticed by the majority of 'gourmet' writers and columnists, whose noses are so deep in the trough of big-business hospitality that they cannot smell a rat when it bites them.

Perhaps our general lack of discrimination is explained by some of the factors mentioned above; but whether or not this is so, I am certain there are defects in flavour with virtually all factory farmed produce.

Taints of fish and of dung are very common, as are peculiarities of texture, and curious tastes possibly associated with the use of hormones; but a generally anaemic *lack* of proper flavour is almost as serious. The extreme lengths some restaurateurs go to to get top quality raw ingredients are not just publicity gimmicks: in the majority of cases they can almost certainly taste the difference, however much the National Farmers' Union or the Meat and Livestock Commission deny that this is possible. And it is not just experts who know a good thing when they taste one – through our business we have learnt that the unsophisticated crofter can often be just as perceptive of quality as the smartest gourmet. If only we could forget the advertisers' promises, the demands of fashion, and our inevitable anxiety not to be cheated, and if we could just relax ... then we might have more confidence in our own taste buds, and discover that on the whole the more 'natural' foods really do taste better.

Outside the subjective realm of taste, where the human tongue is a thousand times more sensitive than any scientific instrument, and the human brain a thousand times more deceiving, there are a number of more objectively quantifiable features of food from intensive livestock farms which have been causing increasing concern. As with the problem of adulteration already discussed in chapter 4, these measurable differences are not necessarily specifically connected with factory farming, but may result from the use of modern methods throughout the whole food production industry. There is a spectrum which ranges from entirely natural food from unpolluted areas of the wild, through old-fashioned 'organic' agriculture to a variety of more intensive methods, and culminating in the total confinement systems of the factory farm. Correlated loosely with this spectrum are certain nutritional defects and chemical or bacterial residues which, while they are not unique to intensively reared animal products, are an inevitable feature of them.

To a large extent, knowledge of these differences is of undesirable extras that are in factory farmed food, rather than of elements which such food lacks. But one or two pieces of research which have concentrated on the nutritional value of food from different livestock systems have indicated that there are analysable variations. Work at the Nuffield Institute of Comparative Medicine has shown that the fat of free range pigs contains a higher proportion of polyunsaturates than that of intensively reared animals (25 per cent of the total fat compared with 17 per cent). The difference may be related to the amount of exercise taken by the pigs, or it may be a function of diet. In a small scale trial by Ann Petch, traditional breeds of pig were fed rations containing from 10 to

60 per cent of grass. Those on the high grass feeds put on weight rather slower, but produced leaner carcasses with twice the proportion of nutritionally desirable polyunsaturated fats, and a slightly improved flavour. Until the right way can be found to promote 'unsaturated' pig meat, the slower growth rate will prevent it competing directly with meat from animals fed the normal wheat and barley rations.[2] Although unsaturated fats are desirable from the nutritional point of view, they are viewed with suspicion by the meat trade, where firm, white (highly saturated) fat has long been regarded as desirable. Crawford points out that the meat eaten by the Masai, among whom heart attacks are virtually unknown, is very different from that which comes from our intensively fed and underexercised farm animals. The proportion of polyunsaturates in the fat of nomadic beef is 48 per cent compared with only 8 per cent for intensively produced carcasses. He describes the 'marbling' of white fat prized by butchers as 'pathological', and asks whether it can really be healthy for us to eat animals which themselves have so unnatural a life style.[3] Studies of the quality of eggs produced under battery, deep litter and free range conditions showed that, on the same diet, free range hens laid eggs containing 70 per cent more vitamin B_{12} than battery eggs, and 50 per cent more folic acid; though in most other physical and chemical respects there were no significant differences attributable to husbandry methods.[4] At the time this research was carried out, these differences seemed unimportant, but recently folic acid deficiency in the national diet has been causing some concern (see page 188).

The research on nutritional composition of eggs showed that feedstuff quality had a significant effect on a wide range of the compounds analysed for. While much work has been done towards optimizing feeding materials, the general emphasis has tended to be on weight gain per unit cost, rather than on nutritional quality of the end-product. No agricultural by-product is so disgusting that it cannot be incorporated somehow in a 'balanced diet' for some unfortunate livestock. Dried poultry manure, ground bones, corn husks, peanut shells, pineapple tops, hydrolysed feathers – the list is practically endless. Even shredded cardboard has been tried as a component of beef cattle's diet. Recycling urine and faeces for the animals to drink and eat a second time is a feature of many intensive livestock units; according to *Hog Farm Management* magazine, one American pig farmer economizes by leaving pregnant sows to forage in the manure waste pits, which is their only source of food for up to 90 days.[5] At Trawsgoed Experimental Husbandry farm, highly polluting silage effluent is being treated with formalin and fed to cattle at rates of up to 40 litres (8 gallons) per day.[6]

While one can applaud the frugality and indeed the ingenuity of such expedients, the question remains whether nutritional science can really formulate wholly adequate, healthy and safe diets based on constituents so far removed from the animals' natural preferences.

From the animal health angle, inadequacies in proprietary or home-formulated diets may be more serious for intensively kept animals, which do not have the option of foraging to compensate for deficiencies in feeding. Certainly, processed commercial pet foods do not always suit dogs and cats, and have been implicated in a number of diseases. Zinc absorption in dogs is sometimes blocked by pet foods with a high cereal content, resulting in stunted growth or skin disease; while too much bone meal can induce thyroid disorders, and excess protein is linked with kidney problems. High levels of magnesium in pet foods have been held responsible for cystitis and urinary blockage in cats.[7] Most proprietary brands of cat and dog food involve a compromise between quality and price, with the temptation for the manufacturer to include inferior low cost ingredients. When these require sterilization to make them safe, the nutritional quality suffers further, as the heat treatment breaks down essential amino acids and enzymes. Farm animals are normally slaughtered quite young, before they show long-term signs of deficiencies, but the evidence from domestic pets indicates some of the difficulties in formulating a satisfactory diet. In many cases, inadequate or incorrectly compounded feedstuffs must adversely affect both the health of livestock, and the nutritional quality of products derived from them.

For people who have a fair choice of foods, inadequacies in one food product can often be balanced by taking more of another, and deficiency diseases are comparatively rare in the affluent nations. What worries shoppers in the supermarket is not that their diet may lack essentials, but that it includes a great many extra constituents of unknown safety. With the majority of these contaminants there may be little direct evidence of any danger – indeed, virtually all have been passed as safe on the evidence of animal testing – but in a disturbing number of cases there is circumstantial evidence which casts doubt on the validity of their clean bill of health. It must be remembered that even the most expert opinions on safety are inevitably provisional until new substances have been in use for a long time: the proof of the toxicological pudding is very much in the eating.

Only a minority of the chemicals found in fresh animal produce before processing have been introduced in a deliberate attempt to alter the product quality and make it more attractive to the consumer. The most notorious is probably the orange dye which is added to the feed of battery hens to colour the yolks of their eggs. Unless hens have access to grass, their eggs have unappetizing pale yellow yolks; and while it is possible to improve yolk colour by adding chopped grass to their rations, it is more convenient to add a synthetic carotin pigment to the feeding stuff. Canthaxanthin, produced by the Swiss chemical firm Hoffmann LaRoche, has been widely used in Britain, though other European countries have banned it on safety grounds. An EC directive forbidding its use in animal feedstuffs was withdrawn after the British government produced scientific evidence for the safety of canthaxanthin (evidence, incidentally, which was all supplied by Hoffmann LaRoche). In the *Final Report on the Review of the Colouring Matter in Food Regulations 1973*, produced by the Food Advisory Committee in 1987, it appeared that the Ministry of Agriculture were having second thoughts; on the grounds that canthaxanthin has been shown to affect the retina, impairing twilight vision and dark adaptation time, and increasing sensitivity to glare. The Food Advisory Committee also recommended that eggs coloured by an alternative pigment, citranaxanthin (also banned as an additive to food for human consumption) should be 'labelled appropriately'.[8]

Canthaxanthin has also been used extensively in fish farming as a feed additive, but doubts over its safety have led to its replacement by the alternative colourant astaxanthin, which is claimed to be chemically identical with the natural pigmentation of wild salmon.[9] The pink flesh of wild salmon derives its colour, and probably also its flavour, from the fishes' natural diet of small crustaceans. Since nobody would want salmon of the pasty white colour which would result from the normal diet given to farmed fish, some sort of artificial colourant is essential for the industry's commercial credibility.

Chemical methods are also used to 'improve' the quality of beef carcasses. More than 10 per cent of cattle slaughtered in Britain in 1989 received an injection of papain, a protein-digesting enzyme derived from the pawpaw, about half an hour before they were killed. This has a remarkable effect on the resulting meat: it can make 'the toughest old cow as tender as the most expensive rump steak'. In fact, the tenderizing effect is so dramatic that the liver and kidneys of treated cattle cannot be sold for normal use, because they disintegrate on cooking. The process was developed by Swifts of Chicago, a major meat wholesaler, who market the resulting meat under the label Pro Ten beef. At first the practice was restricted to elderly dairy cows, but it has increasingly been used to obtain more roast-quality cuts from prime beef cattle. The process was criticized by the Farm Animal Welfare Council, as the injection of 100 ml of enzyme into the jugular vein sometimes causes the cow to collapse with shock, or even drop dead. The uncooked meat contains about 4 parts per million of papain, which is activated and largely broken down during cooking.[10,11] As far as health hazards from meat treated in this way are concerned, there is no reliable evidence, but in November 1989 David Maclean, Britain's Minister for Food Safety, announced that the practice was to be banned in compliance with draft European proposals forbidding the marketing of enzyme-treated meat after 1992.

Biochemical methods of adjusting the carcass quality of meat animals may well become increasingly significant. The demand for leaner meat has stimulated research into ways of reducing the amount of fat on carcasses. One way of increasing growth rates, and thus allowing animals to be killed younger and leaner, is by the use of hormones of the steroid type, or the more recently developed recombinant bovine growth hormone, BST. Alternatively, medication with anti-obesity drugs of the type now under development for humans could result in leaner animals at the time of slaughter.[12,13] The illicit use of these 'β-blockers' to improve lean : fat ratios is already widespread in Holland.[14] Any of these techniques, even if commercially viable, will inevitably leave residues in the carcass, against which considerable consumer resistance can be expected. A third way of producing leaner meat is being investigated at the Hannah Research Institute in Scotland. By immunizing young rats with antibodies to their own fat cells, it is possible to cause the breakdown of up to half the fat cells, producing extra protein and water. As the animals grow, fat deposition normally proceeds by the existing fat cells becoming larger, rather than by the production of new cells. Thus the immunized rats put on far less fat than normal, even after the other physiological effects of immunization have worn off.[15] The advantage of using the immune system in this way is that within a few weeks of treatment there are no detectable residues – in fact it is as if some of the fat had simply been surgically removed.

The vast majority of synthetic chemicals which are given to farm livestock are not intended to affect the quality of their milk, meat or eggs, but are directed at increasing yields, or maintaining the animals in an adequate state of health. The synthetic oestrogen diethylstilboestrol (DES) was widely used in Europe and the United States before being banned as a possible cause of cervical cancer; and the male hormone testosterone, which was used in Britain as a growth promoter until 1986, has recently been implicated as a possible cause of higher than normal cancer rates among butchers and meat-cutters.[16] The EC countries have now decided to ban all growth stimulating hormones, but a number of similar products are still widely used in the United States, and there is considerable pressure for their re-introduction on the other side of the Atlantic. According to the *Meat Trades Journal*, there is a substantial black market trade in steroid growth promoters, which are still in regular use on many farms.[17]

In addition to their possible role as carcinogens, the use of such hormones may have more bizarre side effects. A survey conducted by workers at the Dundee Medical School showed that in the 1970s, when female hormones were used as growth promoters in cattle, 98 sons were born to butchers for every 100 daughters (the national average is 106 boys per 100 girls). In the early 1980s, following a switch by farmers to predominantly male hormones, the ratio of children born to butchers was 121 sons to every 100 daughters.[18] While these statistics are merely amusing, the experience of thousands of Puerto Rican children borders on the tragic. In a syndrome known as thelarche, young girls under eight, and in some cases only two or three years old, develop enlarged breasts and pubic hair, and commence menstruation. An article in the *New England Journal of Medicine* connected a recent epidemic of thelarche in Puerto Rico with high rates of consumption of hormone-treated chickens.[19] While not all epidemiologists accept this claim, similar experiences in Italy were partly responsible for the 1988 ban on all hormone growth promoters in EC countries.

The antibiotics and other growth promoters which many farm animals receive on a routine basis are responsible for significant residues in animal produce. To improve the efficient conversion of feeding stuff into body tissues, a number of feed additives are used. The object in each case is to encourage beneficial micro-organisms in the animal's intestine, and to discourage those bacteria which do not assist growth. In addition, antibiotics appear to reduce the gut wall thickness, which increases the rate of nutrient absorption. The antibiotics used for this purpose in the United Kingdom are a different group from those used in the treatment of diseases in animals or humans, and have been chosen on the basis that they tend to remain in the gut rather than being absorbed into the bloodstream. Because of this, and because the routine doses are small, residues of these compounds in meat are relatively low, and in the majority of cases there is no legal requirement to withdraw them for a period before slaughter.

In the United States and many other countries, such a division does not apply, and antibiotics like penicillin and tetracycline (banned in Britain as routine feed additives) are used indiscriminately for therapeutic and non-therapeutic purposes. Doses of 2 to 50 g per ton of feed can be used for improving feed conversion, or up to 1000 g per ton as a protection against disease. Significant levels of antibiotics are found in many carcasses, even though medication should be stopped for a period immediately before slaughter. Neomycin is claimed to account for a quarter of all chemical residues detected in cattle carcasses in the United States.[20]

In fact, the British distinction between routine feed drugs, available without prescription, and therapeutic drugs for which a vet's prescription is required, is more apparent than real. In the overcrowded conditions of intensive livestock houses, widespread use of therapeutic drugs is practically the norm. Even if only a few members of a herd contract an infectious disease, normal veterinary practice is to 'treat the whole herd as an individual', in other words to dose the animals as if they were all sick. For the control of enzootic pneumonia, a disease which affects a high proportion of bacon-weight pigs, the recommended treatment at the first sign of coughing is in-feed medication with tetracycline at 300–600 g per ton.[21] The following comments by the editor of the *Veterinary Record* on the use of antibiotics in calf rearing might equally apply to other forms of intensive livestock husbandry: 'The emphasis has shifted from their use as growth promoters to prophylaxis to counter the stress imposed on calves by modern practices in the calf industry. Consequently higher levels are being used over unnecessarily long periods of time. Instead of being reserved for proper therapy, antibiotics are being administered increasingly to counter unsatisfactory husbandry practices'.[22]

Quite simply, more intensive farming methods lead to greater need for antibiotics, and the statistics for antibiotic consumption reflect the general trend towards larger and more intensive units. In Sweden, the use of antibiotics in pig feeds has been banned, but there has been a corresponding decrease in growth rates, and a massive increase in diseases such as swine dysentery.[23] A major cause for concern over the use of antibiotics is that the more antibiotics are used, the greater the chance of disease organisms developing which are resistant to antibiotics. This is already a problem in the British calf industry, where a strain of salmonella resistant to at least five major antibiotics has become endemic.[24] The possible risks to human health from this development are discussed below.

Numerous drugs apart from antibiotics are used in farming, from anti-stress sedatives to appetite boosters, and many of these can result in contamination of produce. Farmers are required to observe minimum withdrawal periods before slaughter, but not all do so. In 1985 more than a quarter of the pig kidneys sampled by the Ministry of Agriculture contained up to eleven times the permitted residue level of sulphadimidine, an antibacterial drug used to treat rhinitis. In the United States, there are proposed bans on sulphadimidine (sulfamethazine), and the pig growth promoter carbadox, because they are carcinogenic; while chloramphenicol, another widely used livestock antibiotic, has already been outlawed because of a suspected link with leukaemia.[25] Detectable residues are not necessarily harmful: for example, the 1 part per million of arsenic permitted in poultry meat is about a quarter of the level occurring naturally in fish, while shellfish can contain up to 100 times as much. With novel compounds, the assessment of safe residue levels presents great difficulties, and the scale of operation required is daunting. In 1985 it was estimated that 20,000 different brands of animal drugs were in current use on American farms; and the General Accounting Office of the United States has produced a list of over 1000 substances used as drugs or pesticides in livestock production, of which 140 are likely to occur as contaminants of animal produce.[26]

Many of these foreign substances are ingested incidentally by the animals as more or less accidental ingredients of feeding stuffs; but at least one insecticide is deliberately fed to poultry as the most efficient way of directing it to its intended target. The wire floors of battery cages separate the hens from their droppings in what appears to be a most hygienic manner, and indeed the system is effective in preventing coccidiosis, a parasitic disease of the gut which is transmitted from bird to bird via the faeces. However, battery units are much more vulnerable to fly infestation than houses where the birds are kept on litter, since the birds cannot peck up the maggots which feed on their droppings. As well as being a nuisance, flies can transmit avian flu virus from bird to bird. It is possible to keep the entire battery house doused in an insecticide-laden mist, but the manufacturers of Larvadex came up with an ingenious alternative. This insecticide is mixed with the hens' food, passes through the birds into their excreta, and kills the maggots then and there as they hatch out. Larvadex was the subject of controversy almost as soon as it began to be used in the United States. In 1983 it was temporarily withdrawn, after accusations that the eggs and meat of poultry that had been dosed with Larvadex contained traces of a carcinogen called melanine. In June 1984 the US Environmental Protection Agency (EPA) was embarrassed by the disclosure that teratogenic toxicity tests showed that Larvadex caused damage to foetuses, even at the smallest doses tried. John Moore, head of the EPA pesticide division admitted that somehow a review of these findings 'didn't get done. It got lost'. The product was immediately banned again. Undaunted, the manufacturers commissioned further toxicity studies, which came up with the welcome news that residue limits of 2.5 parts per million in eggs could be counted perfectly safe.[27] In 1985, Larvadex was given the all-clear in the United States, and this ingenious product will no doubt soon find its way to the rest of the world.

Agricultural chemicals which are not intended to be ingested can also find their way into body tissues. The use of lindane in sheep dips was recently banned in Britain after residues were discovered in lamb meat. Many pesticides and other toxins find their way into the diet of farm animals unintentionally. When livestock are kept permanently confined and unable to forage, they are more or less obliged to take the food and water on offer, however unpalatable. The restricted diet of intensively housed livestock can, unless carefully supervised, lead to the animals consuming toxic substances which may impair reproduction, reduce efficiency of feed conversion, cause disease, or leave residues harmful to people. It is thus in everyone's interest to try to keep such 'xenobiotics' out of the food chain, but unfortunately this is very difficult in practice. Leaving aside such accidents as the mixing of a polybrominated biphenyl (PBB) instead of a mineral supplement into a batch of animal feed in Michigan in 1975 (with the result that thousands of animals had to be destroyed, and local milk supplies were contaminated with dioxins), poisonous and persistent chemicals are so widely used in agriculture that groundwater and crops almost inevitably contain unwelcome traces of toxins. The processing industry is partly responsible for the general overuse of insecticides. Manufacturers have learnt that while the public may put up with the odd caterpillar or beetle in produce fresh from the market stall, they get very upset if these creatures turn up in a can or freezer-pack bearing a famous brand name. Packagers therefore set very high standards regarding pests and blemishes, and it is much cheaper for producers to make a few extra applications of pesticide than to lose money by having their crop 'downgraded'.

A study by the National Research Council in America produced the following list of the fifteen foods most likely to contain carcinogenic residues (in decreasing order of toxicity): tomatoes, beef, potatoes, oranges, lettuce, apples, peaches, pork, wheat, soyabeans, carrots, chicken, maize and grapes.[28] The problem of xenobiotic residues is obviously not restricted to factory farmed produce, but extends across the whole range of most people's diet. In general, though, organic vegetables and free range animal produce should be much less contaminated than the products of more 'advanced' agriculture.

Animal feedstuffs are frequently compounded from crops and meat by-products that have been condemned as unfit for human consumption, and this can lead to higher than normal levels of xenobiotics. In 1982 milk in Hawaii was found to contain nearly ten times the maximum allowable level of heptachlor after cattle were fed pineapple tops which had been treated with this pesticide.[29] And contaminated grain in cattle feed recently led to milk from over 1000 British farms being condemned because of its high lead content. In many plants and animals poisonous materials can build up in tissues to levels many times greater than the original levels in the environment. Shellfish are the classic case, concentrating both industrial pollutants such as heavy metals, and natural toxins from seasonal algal 'blooms', sometimes to levels which cause acute poisoning in people who eat them. Within the body of an animal, toxins are partitioned very unequally between different tissues. The liver and kidneys frequently retain xenobiotics in high concentrations; and even if such organs are condemned as human food, there is a chance they will find their way back into animal feedstuffs and the pollutants will be recycled.[30] The same potential danger attends the use of dried animal faeces for feeding livestock. Since many of the most toxic pollutants are fat-soluble, they may accumulate to high levels in fatty tissues, and in egg yolk and milk fat. Aflatoxins from mouldy feeds, pesticides such as DDT (dichloro diphenyl trichloroethane), and industrial pollutants such as dioxins and polychlorinated biphenyls (PCBs) have all found their way into cows' milk. Similar partitioning can also occur in humans, of course, and a 1986 study by the British Working Party on Pesticide Residues in Food noted that the DDT intake for some breast-fed babies was higher than the World Health Organisation's acceptable daily limit for adults.[31] More recently, pregnant and lactating women were warned that drinking two pints of cows' milk per day could lead to harmful quantities of dioxins being passed on to their babies.[32]

There is some evidence that residues bound in animal tissues are less toxic than before such binding, which would imply that the danger of residues in meat may have been overstated. In one study, rats were fed on the livers of other rats which had eaten aflatoxins labelled by radioactive tracer methods. The metabolized aflatoxins appeared to be much less readily absorbed, and the liver-fed rats were less inclined to get cancer than those fed an equivalent quantity of unbound aflatoxin.[33] It is difficult to be oversanguine on the basis of such experiments: the established impact of DDT on birds of prey suggests that it may be wiser to expect the unexpected.

How does factory farming fit into the worldwide problem of exposure to environmental pollution? Obviously modern farming methods contribute to the volume of pollutants, as will be discussed later; but the degree of risk from chemical residues in farm produce is more a function of feeding and cropping methods than specifically related to the way in which livestock are kept. Eggs, milk and meat from intensively kept animals will tend to contain higher levels of antibiotics, and, because the animals are often fed on contaminated feedstuffs, more pesticide residues. Free range animals are not entirely free from risk: grazing animals can drink from polluted water sources or eat poisonous leaves and berries. The pasture itself may be contaminated by industrial chemicals even if it has not been sprayed. On rare occasions, for example after Chernobyl, the diet of housed livestock might even be safer. As *Poultry World* crowed triumphantly in response to government advice on coping with nuclear war, 'Battery Birds Better Off After the Bomb!'

6
In Sickness and In Health

The most immediate human health hazards connected with factory farming are the result of biological, rather than chemical, contamination. Battery chicken and salmonella now go together as inevitably as love and marriage in the old song – though a junior health minister in the British government was recently sacked for saying as much in public. But isn't food poisoning just a bit of a joke, like the common cold? Is it really worth knocking the farmers just because people are occasionally sick after eating badly cooked chicken or egg dishes? Unfortunately, it appears that meat and egg-borne diseases are increasing both in frequency and in severity, and soon they may not seem so funny at all.

Animal products such as meat and milk offer the ideal growth medium for many bacteria, and contamination can lead to the presence of high concentrations of potentially harmful organisms. About three-quarters of the established causes of food poisoning are meat products of some sort.[1] Pathogenic bacteria in food may cause infection of the consumer with diseases from cholera and tuberculosis to typhoid fever and various relatively mild forms of 'food poisoning'. Some bacteria are not directly harmful when eaten, but produce toxic substances as they multiply in contaminated foodstuffs: this is the mechanism of staphylococcal food poisoning and the more serious botulism. Correct handling of foodstuffs can greatly reduce the chances of food poisoning. Since thorough cooking kills the bacteria responsible, foods eaten raw or lightly cooked, and those kept for prolonged periods after cooking are the major risks.

However much care is taken in handling food in the kitchen, the risk of bacterial infection will inevitably depend on the initial degree of contamination, which can be from a variety of sources. Many of the common food poisoning bacteria thrive both in man and in other species, including food animals. So while infected food handlers are the main cause of poisoning related to bacteria of the *Staphylococcus aureus* group, which flourish particularly in cured meat products, the same type of poisoning can also result from eating cheese which has been infected with staphylococci from cows, through the use of unpasteurized milk. Most causes of salmonella poisoning, on the other hand, are probably started by meat from infected animals, though human carriers are responsible for some outbreaks, and others are initiated by bacteria carried by rats, birds or domestic pets. (Figure 2 shows the incidence of salmonella poisoning in England and Wales between 1941 and 1984.)

Because of the variety of routes by which bacterial contamination occurs, it is impossible that we shall ever get rid of food poisoning completely. The fact remains, though, that bacterial food poisoning has increased substantially in the past forty years, and there are probably two main factors which account for this trend. The first is the increased scope for cross-contamination of food products with the modern tendency towards centralized processing and preparation on a large scale, so that each original source of infection can affect an ever-increasing number of consumers; and the second is an apparent increase in the infection of live food animals with pathogenic bacteria, so that more eggs and carcasses are contaminated right from the start.

The most striking cases of food poisoning occur in schools, hospitals and other institutions where the scale of catering means that each incident tends to affect a large number of people. But the statistics show an increasing number of outbreaks, not just an increasing number of victims; and this should hardly be affected by the size of group being fed at one time. Centralized processing can give rise to numerous outbreaks of food poisoning which are not apparently related; and although modern food factories are on the whole very hygienic, when mistakes do occur they can be disastrous. The most widespread instance of salmonella poisoning from a single source occurred in 1984 in the United States, where between 150,000 and 200,000 people were affected by milk from a dairy in Illinois. No faults were found in the pasteurizing system, and the source of contamination was never discovered.[2] In Britain the following year sixty-three infants were infected with a rare strain of salmonella after consuming dried milk powder from Farley's Kendal factory. Although the number of cases was small, confidence in the company was so shaken that within three months it was on the brink of liquidation. When the Kendal plant was bought by Boots the Chemists, they spent £900,000 cleaning the contaminated vacuum hopper which had eventually been traced as the source of the infection, but after 15 months they had still failed to sterilize it effectively, and decided to build a completely new plant away from the old site.[3]

Techniques such as pasteurization and sterilization of milk have been immensely helpful in reducing food-borne disease, and they obviously

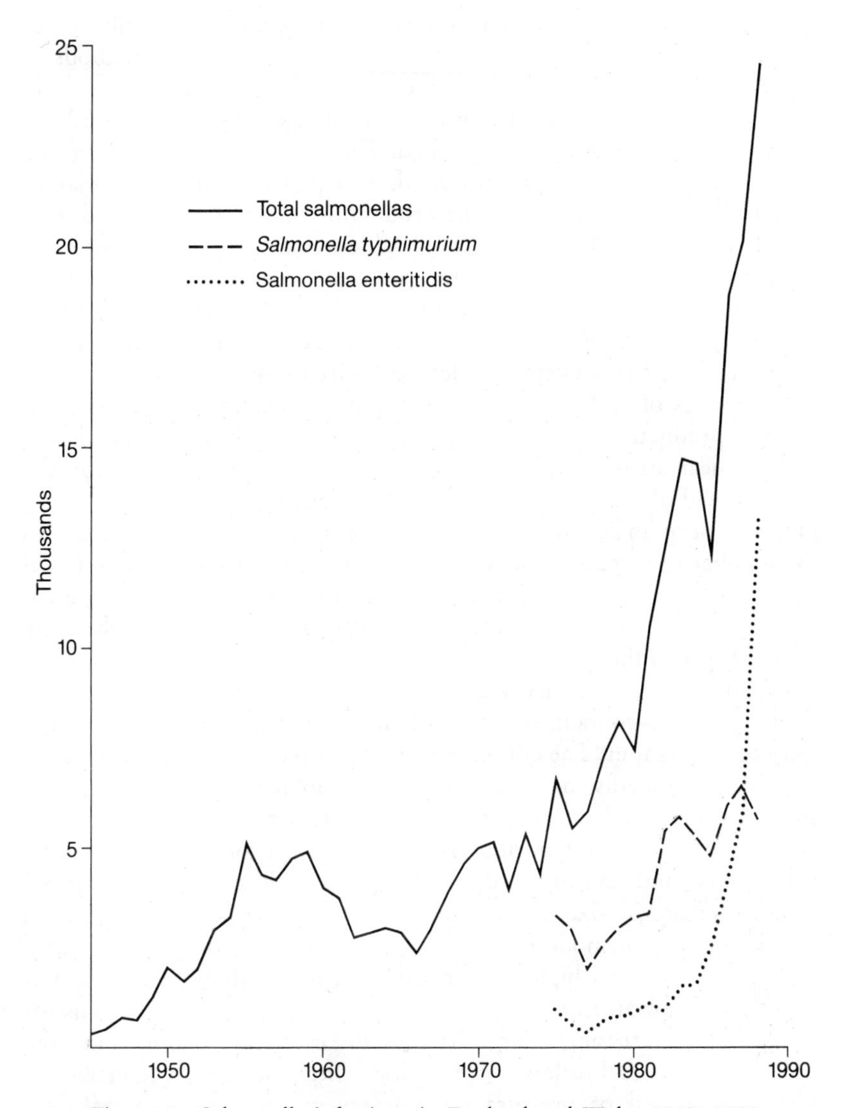

Figure 2 Salmonella infections in England and Wales 1945–1988. (Adapted from data in London Food Commission *Food Adulteration* and from OPCS *Monitor: Infectious Diseases*.)

require a degree of centralization to obtain enough throughput to justify the expensive plant involved. In other areas of food processing, however, the application of production line techniques can actually lower standards of hygiene compared with the more old-fashioned methods of batch production. If a single element in a continuous process food factory becomes infected, there is a high probability that all the products from the factory will be contaminated. The economic advantages of one large plant over ten smaller ones can thus be partly negated by the larger scale of the public risk should the production line become infected with harmful bacteria. An investigation which was carried out into the frequency of salmonella in pork sausages supports this hypothesis. None of the sausages from small manufacturers were found to be contaminated, whereas of those from one large manufacturer, nearly 50 per cent of the packets sampled harboured salmonella.[4]

Economies of scale in livestock markets and abattoirs are even more counter-productive in public health terms. As beasts from larger geographical areas mingle at auction marts, so opportunities for the rapid spread of infection increase. And as slaughterhouses handle larger numbers of animals, and go over to more production line methods, so the number of carcasses contaminated from a given bacterial source will tend to become larger. In 'line processing' the animals are stunned, bled, eviscerated, skinned and cleaned by different operators as they are moved through the plant mechanically on an overhead conveyor system. Each operator handles many carcasses using the same tools and equipment, and cross-contamination is difficult to control, particularly at high rates of throughput. The conditions in British slaughterhouses have been criticized repeatedly by EC inspectors, who have claimed that they show a general lack of awareness of basic hygiene, which allows gross contamination of meat during carcass dressing. Production lines often run too fast, and staff move too freely between 'clean' and 'dirty' areas.[5]

The increase in size of slaughterhouses has been accompanied by a certain amount of modernization, and the potential for greater control of hygiene, but in a highly competitive business this potential has not been fully realized. Intensive livestock units tend to produce animals for slaughter in increasingly large batches for economic reasons; and the arrival of 100,000 broilers more or less simultaneously will inevitably put a good deal of pressure on workers in even the most efficient purpose-built poultry abattoir. The highly mechanized system of stunning, scalding and plucking, cooling in large baths of iced water, and final packing or freezing means the birds need not be touched by human hands after they are shackled on to the conveyor belt on arrival, but nevertheless the opportunities for cross-contamination are legion.

Three-quarters or more of the chicken carcasses which reach the shops are infected with salmonella, and although stringent hygiene in the kitchen may stop any of the bacteria being transferred to work surfaces, knives and other food items, this is hardly a good start.[6] Not *all* the blame for food poisoning can be attached to the careless housewife or restaurateur.

Nor, in fact, is the meat processing industry the sole villain of the piece: modern farming methods are probably the original source of most of the bacterial contamination of meat, and if salmonella is found in eggs, there is no way it could have got there after leaving the farm. Many farm animals are infected with food poisoning bacteria – just how many is difficult to estimate, as few random surveys have been conducted, and while acute cases of salmonellosis should be reported, many animals may harbour the disease without being apparently sick. The crowded conditions of factory farms favour infection of one animal from another, through flocks of ever-increasing size. The most common way this happens is probably by animals eating faecally contaminated food, but even where this cannot happen, the disease organisms are transmissible in the air, particularly in a warm moist atmosphere.[7] Just as human beings continue to excrete infective organisms for weeks after an illness (or even for years in some cases, such as the typhoid bacillus, *Salmonella typhi*), so many farm animals pass such bacteria in their faeces. Again as with human beings, infections are readily transmitted in surroundings that are overcrowded or insanitary; and the more fluid the faeces, the greater the risk.

Despite the best efforts of farmers to sterilize buildings between different batches of stock, infections such as salmonella inevitably seem to recur, and while sometimes people or wild animals may be the vehicles of infection, the most common cause is believed to be contaminated feeding stuff. (The influence of feeding habits on salmonella excretion has been studied in pet animals, and also in wild birds – those feeding on tips and near sewage outflows often being carriers, while country birds in sparsely populated areas are seldom infected.) Intensively kept farm animals are invariably fed high protein feeding mixtures which often contain ground up slaughterhouse waste and other animal by-products such as dried poultry manure. The inclusion of heads, feet, internal organs, bones and skins from infected animals, along with their droppings, in feedstuffs for the next generation, is a sure recipe for perpetuating diseases, unless proper sterilization procedures are rigorously adhered to. The heat treatment required is expensive, and inevitably lowers the nutritional content of the feeding materials somewhat, so there is a temptation to the producer to cut corners. But even with

The meat trade creates huge quantities of offals, which are rendered down at high temperature and form an important ingredient of much animal feeding stuff. There is a danger that infections will also be recycled, or (as in the case of BSE) may pass from one species to another. (The *Independent*)

the best will in the world, with 'clean' and 'dirty' materials on one site it may be very difficult to prevent cross-contamination of the sort that occurred at the Farley's milk plant. Recently, the feeding of poultry offal and feather meal to laying hens, and of recycled animal products to milk cattle, have been prohibited in the United Kingdom.

In 1986 more than a quarter of the samples taken from producers of processed animal protein were found to contain salmonella, and thus fail to comply with the UK Diseases of Animals (Protein Processing) Order of 1980.[8] The situation had not improved by 1988, when it was highlighted by a national scare about salmonella in eggs, but despite their failure to comply with the regulations, none of the millers had ever been prosecuted. Under regulations introduced in October 1989, protein processors have to submit weekly samples for testing, and are forbidden to release any contaminated products; so perhaps some improvement in the situation can be expected.

The number of reported cases of salmonella poisoning has risen steadily in recent years. In 1988 there were 8817 people affected in the three months ending on 30 September; in 1985 there had been just 4447 cases, and before 1970 the average for the same quarter was less than

2000 affected individuals.[9] The dominant strain until recently was *Salmonella typhimurium*, and conventional wisdom held that on the occasions when eggs were the vehicle of infection, the germs must have been carried on the outside of the shell. This theory was disproved as more cases of food poisoning began to be caused by *Salmonella enteritidis*, which has been shown to thrive in the hen's oviduct, and can be incorporated in the egg before it is laid. There were 975 isolations of *S. enteritidis* in the autumn quarter of 1985; by 1988 this strain was responsible for 5262 cases of food poisoning in the corresponding quarter, with eggs frequently implicated as the source of infection.[10]

Following Mrs Edwina Currie's exposé of the scale of egg-borne salmonella, the British government started to test all laying flocks for the presence of salmonella, and to slaughter all flocks containing infected birds. The proportion of eggs actually containing traces of salmonella has been estimated as about one in 7000, with only about one in 70,000 containing enough bacteria to cause food poisoning if not properly cooked.[11] During 1989 half a million hens were slaughtered, new restrictions on permitted food ingredients were announced, and routine testing of all commercial laying flocks was instituted. The testing regime announced by the government bore heavily on the smallest producers. To be 95 per cent certain that tests were showing up infected flocks, a battery farmer with 20,000 birds had to submit cloacal swabs for testing from sixty of them; a small farmer with 100 birds had to have fifty of them tested regularly, while someone with just twenty must have them all tested.[12] After anguished protests from small producers, the procedure was modified so that droppings could be provided for analysis instead – cutting the cost from over £120 to around £30 for the laying period.[13] These regulations are obviously an improvement, but in view of the minuscule fraction of total UK egg production from flocks of under 100 birds, it might be more realistic to exempt such small producers altogether.

Salmonella is not the only disease contracted from food, though it is the most commonly reported and fastest growing source of food poisoning in Britain, accounting for 90 per cent of the 21,000 incidents recorded in 1987, and causing round about fifty deaths per year, mostly among infants or the elderly. In 1977, a link was established between the organism *Campylobacter* and food poisoning-type symptoms, and by 1981 more cases of gastro-enteritis were attributed to this bacterium than to salmonella, though it has not yet found its way into the official food poisoning statistics.[14] Circumstantial evidence links *Campylobacter* poisoning strongly with the consumption of poultry meat, nearly all of which is infected with the organism. *Campylobacter* flourishes at very low temperatures, and refrigerated meat counters may thus provide it with the ideal environmental 'niche'.[15] The deadly *Clostridium botulinum* can also affect farm animals, and thus find its way into human food. In a recent case where cattle were being fed on ensiled poultry litter, 80 out of 150 cows developed botulism, and 68 died. *Clostridium botulinum* type C organisms were present in kidneys from the same herd, and meat which had been frozen for several months contained appreciable residues of type C1 toxin.[16]

The claim has been made that farmers are unwilling to take any action to control food poisoning organisms in their animals, because unlike such food-borne diseases as tuberculosis or brucellosis, salmonella and similar bacteria do not have a major impact on production.[17] This is particularly the case with *Salmonella enteritidis*, which can infect egg yolks within the hen's oviduct without producing obvious symptoms or reducing yield. In broilers and more particularly in calves, salmonella infections reduce productivity quite drastically unless treated with antibiotics. Despite the level of publicity it has received, however, the problem of rapidly increasing rates of salmonellosis can hardly yet be compared in effect with the thousands of deaths each year from tuberculosis in the first half of this century. Also, with so many links in the chain of infection, it is a little unfair to expect action from individual farmers, or even from the farming industry as a whole, when what is needed is an effective national policy regulated by government. The eradication of cholera, typhoid and tuberculosis by ensuring the safety of drinking water and milk supplies are major achievements of the past 150 years, but we should not be too confident that serious food-borne disease is a thing of the past. To expect healthy food from unhealthy farm animals may be asking too much, and despite the efficient factory farmer's concern with hygiene and liberal use of medication, it is far from certain that the general health of farm livestock is improving.

Listeriosis is a serious disease which appears to be increasing, both in farm animals and in the human population. The micro-organism *Listeria monocytogenes* causes septicaemia, encephalitis and abortion in sheep and cattle, and more rarely produces similar effects in human beings. It is only recently that epidemics of listeriosis in humans have been traced back with reasonable certainty to animal sources. Dairy products (in which the bacterium gains some protection from gastric acids) have been to blame in several cases, though another epidemic was traced back to coleslaw made of cabbages fertilized with manure from infected sheep.[18,19] When it is contracted by pregnant women, listeriosis often results in abortions or neonatal mortality, and it sometimes causes meningitis in adults. Only a hundred or so people are affected per year in the United Kingdom, but the mortality rate is around 35 per cent. There is mounting evidence that silage feeding is responsible for the increasing frequency of this disease: the micro-organism responsible can survive for six or seven years in silage clamps, and contaminated silage has been implicated in a number of outbreaks on farms.[20–2] A recent editorial in *The Lancet* suggested that listeriosis is yet another disease for which infected chicken meat might be responsible, a theory which attracted an angry response from agricultural scientists.[23] In the majority of cases, it has been impossible to demonstrate any association between human listeriosis and the consumption of any kind of food, but the observation that *Listeria monocytogenes* occurs frequently in the faecal bacteria of abattoir workers strengthens the presumption that farm animals are a significant reservoir of infection.[24,25]

Farm workers, slaughtermen and vets are particularly at risk from a number of animal diseases that can be communicated directly. Brucellosis is probably the most familiar occupational hazard, though less common than in the past, but leptospirosis is another serious occupational disease among dairy farmers. Foot and mouth disease, which is uncommon, and anthrax, which is very rare but highly dangerous, can also be transferred to humans, as can the relatively new pig disease of streptococcal meningitis, which sometimes causes permanent deafness.[26] Enzootic abortion in ewes is a well known hazard to pregnant women. Bovine leukaemia is endemic in dairy herds in the United States, and despite general pasteurization, concern has been expressed that the virus may be transmitted to humans.[27] Another cancerous disease called 'leukosis' is common and transmissible among chickens, but it is not clear whether it presents any risk to human consumers of fowl meat.[28]

Grave alarm has been caused recently in Britain by the appearance of a horrifying new disease among dairy cattle. Bovine spongiform encephalopathy (BSE) is a degenerative brain disease which drives cows mad before they lose balance, collapse and finally waste away. Since the early 1980s, meat products have been included in the supplementary feeds given to high yielding dairy cattle, and it is almost certain that the appearance of BSE is a direct result of feeding milk cows on a diet which included imperfectly sterilized infected sheep's brains. It has been claimed that the disease originated from a single batch of cattle feed fed to calves in 1983–4, following changes in the method of sterilization allowed for such feeding stuffs.[29] Scrapie, or 'mad itch', is a fairly widespread disease of sheep, which appears to be closely related to BSE.

In June 1988, on the advice of a working party led by Sir Richard Southwood, Professor of Zoology at Oxford, the Ministry of Agriculture made BSE a notifiable disease, and the following month they issued

Thousands of cattle have been destroyed in an effort to stop the spread of BSE – a brain disease which originated from the use of inadequately sterilized sheep offals in fodder given to dairy cattle. The exact nature of BSE, and the extent of any possible threat to humans, are still uncertain. (© David Jackson)

a temporary ban of the feeding of animal protein to ruminants. By October 1988, fifteen cattle per week were being destroyed, and their carcasses burnt, and the total number of fatal cases of BSE had risen to about 800.[30] There was no immediate evidence that the infected cattle were a human health hazard, but similar degenerative diseases in people had previously been linked with the habit of eating brains. In New Guinea, *kuru* was a common cause of death in cannibal tribes, particularly among the women and children, who were given the brains to eat. A much more common complaint, Creutzfeld–Jakob's disease, occurs among people who eat sheep's brains and eyeballs, such as the Libyan Jews in Israel. The similarities between these diseases were pointed out by Holt and Phillips in the *British Medical Journal*, with the suggestion that clearer guidelines were required regarding the use of brains in human food.[31] In December 1988 the government banned the sale of milk from infected cows, and in June 1989 they banned the use of brains and certain other offals from *all* cattle in food for human consumption.[32] By October 1989, 100 new cases of BSE were being reported each week, and 6831 cows on 4301 farms had contracted the disease; but it was being predicted that BSE would start to decline after 1993–4, and disappear by the year 2000 – if trials confirm that there is no hereditary or contact transmission.[33] Research on scrapie has failed to establish just how the disease is passed on, but there does appear to be a hereditary factor involved, and the infective agent seems to be remarkably resistant to all normal forms of sterilization such as heat and ionizing radiation.[34] BSE is an interesting case of a disease apparently jumping species as a direct result of modern farming practices; and while its implications for human health are not yet apparent, it has certainly been hard luck for the formerly herbivorous cow.

As the process of genetic mutation of viruses to produce new diseases becomes better understood – an area of research which has become more urgent since the appearance of AIDS – there is increasing evidence for the role of cross-infection between humans and other species in the emergence of new mutant viruses. Humans and ducks are susceptible to different types of flu virus, so do not normally infect each other. Pigs, however, can transmit and be infected by both human and avian flu viruses, and it has been proposed that genetic recombination in pigs exposed to both types of virus may be responsible for the production of new human flu strains.[35] This could account for the tendency of major global flu epidemics to start in south-east Asia, where traditional farming methods involve humans living in close contact with pigs and ducks.[36]

A recent article in *Nature* points out that fish farming developments in Asia and Africa may increase human health hazards of this sort.[37]

Fish kept in ponds are often fed on fresh manure from pigs or poultry; and latrines are built over carp ponds, or domestic sewage effluent deliberately channelled into the ponds, to increase the fish's growth rate. (It is reported that the flavour of carp from sewage-rich streams tends to be 'rather disappointing'.)[38] In Thailand an ingenious system is used in which hens are kept in cages above pigs, which are in turn kept in pens above fish ponds; so that the pigs must eat hen droppings, while their own dung goes to feed the fishes. Why anyone would then fancy eating the fish is perhaps difficult to understand, but this homespun variation of factory farming has been enthusiastically recommended by the United Nations Food and Agriculture Organisation.[39] The writers in *Nature*, however, sound a cautionary note: 'The result may well be creation of a considerable potential human health hazard by bringing together the two reservoirs of influenza A viruses, generating risks that have not hitherto been considered in the assessment of the health constraints of integrated fish farming'.

The microbiological effects of modern agriculture may be just as difficult to avoid as its chemical impact on the environment. New viruses spread just as quickly as toxic chemicals diffused in the oceans, and their virulence is not attenuated by distance. As with chemical pollutants, we can avoid some but not all of them by choosing what we eat; but although farm animal diseases are of more immediate danger, and cause more fatalities in the short term, than chemical residues, we tend to accept them with greater equanimity. Illness, after all, is a more familiar experience than poisoning, and doctors are known to be able to help in cases of salmonella or flu, whereas they cannot be relied on to protect us from the random threat of cancers induced by chemical pollutants.

But just as food poisoning cases have increased in number so rapidly over the past few years, so the organisms responsible are showing signs of becoming resistant to medication. The phenomenon of antibiotic resistant strains of bacteria has been recognized for many years, and in 1969 the Swann Report on the *Use of Antibiotics in Animal Husbandry*[40] recommended that as far as possible different antibiotics should be used for treating humans and farm animals. The idea was to avoid the building up in animals (and subsequent transfer to humans) of strains of salmonella resistant to the antibiotics used in hospitals. Although food poisoning is not usually treated with antibiotics, in infrequent cases salmonella bacteria become established in the blood-stream, resulting in septicaemia or meningitis which requires antibiotic treatment.

Unfortunately, Swann's aims have not been achieved: on one hand, reckless prescribing by veterinarians of drugs which were intended to be used only in the gravest cases, and on the other hand a failure of the bacteria to confine their resistance to the feed drugs, have thwarted efforts to stem the tide of more and more highly resistant strains of salmonella. Trimethoprim is one of the best drugs for treatment of typhoid fever and salmonella septicaemia, and in 1979 trimethoprim resistance was rare in salmonellae; but following an intensive promotional campaign to encourage its veterinary use, resistant strains have proliferated. Salmonella resistance to gentamicin, which is used to treat infections such as meningitis, was first reported in Britain in 1982, and has increased considerably since that date.[41] In West Germany, the use of gentamicin to rinse turkey eggs or inject chicks in an attempt to reduce salmonella contamination of turkeys, caused a rapid rise in the number of resistant bacteria.[42] By 1985, three-quarters of all cases of salmonella in calves were caused by multi-resistant strains of the bacteria, and by 1989, multi-resistant strains of *Salmonella typhimurium* from poultry had increased from 2 per cent of all isolates to 7 per cent, though the majority of poultry infections were still drug-sensitive.[43,44] Since cattle and poultry are the major reservoirs of salmonellae from which human infections arise, the development of highly resistant strains through overuse of antibiotics is bad news for us as well as for the calves. In the infrequent cases where salmonella infections result in septicaemia or meningitis, multi-resistant strains are difficult to treat, and patients have died because such infections have not responded to the limited range of drugs remaining available.[45] A significant paper in the *New England Journal of Medicine* in 1984 described a previously unconsidered effect of antibiotic resistance in salmonella.[46] A number of people ate hamburgers mildly contaminated with a multi-resistant strain of *Salmonella newport*, but did not become ill as an immediate result. Only when they took penicillin for other medical conditions, thus apparently killing off the normal gut bacteria which were keeping the *Salmonella newport* strain in check, did they become ill with gastro-enteritis from the proliferating salmonella organisms. If this interpretation is correct, the implications for medicine are serious: the last thing a patient who really needs antibiotics wants is a sudden stomach upset from a flare-up of drug-resistant food poisoning bacteria.

There is still considerable debate over how likely it is that antibiotic resistant bacteria such as salmonellae from animals will transfer their resistance to more specifically human disease germs.[47] While it could happen, the risk of such resistance originating with overuse of antibiotics in human medicine seems much greater. For all the routine medication indulged in by farmers, the average human intake of antibiotics prescribed by doctors is probably five or six times higher per annum than that of farm animals, in proportion to body weight.[48] In a study of antibiotic resistance in human *E. coli*, faecal samples from vegetarians actually showed higher levels of resistant bacteria than meat eaters; on the other hand, it appears that healthy adults working with farm animals have more drug-resistant gut bacteria than their urban counterparts.[49,50] But whatever the uncertainties about means of proliferation of multi-resistant disease organisms, their spread can only be encouraged by the continued overuse of antibiotics by doctors and vets. It is to be hoped that as the mechanism by which disease resistance is transmitted becomes clearer, a more prudent strategy for antibiotic use may eventually emerge.

Antibiotics have failed to solve the problem of food contamination by bacteria; but rather than attacking the problem at its source by abolishing vast overcrowded animal houses and the use of infected feedstuffs, a solution which is now being promoted is the use of irradiation to clean up bacteria laden food products. It has long been known that the ionizing radiations from radioactive elements destroy bacteria, and this effect is used for sterilizing hospital equipment. By subjecting food to controlled doses of ionizing radiation, insect pests and bacteria could be destroyed without significantly affecting the appearance or taste of the food, and without making it measurably radioactive. The main advantage for the producer is an improvement in keeping qualities: delayed ripening of fruit, and a prolonged shelf life for other perishable products. The doses proposed for foodstuffs (about 100 million times more than in a chest X-ray) are not enough for complete sterilization, which requires radiation levels that would substantially change taste, smell and texture; but nevertheless they produce a very marked decrease in levels of bacterial contamination.

For the past thirty-five years the nuclear industry has spent large amounts of time and effort in attempting to prove that food irradiation is safe, to overcome widespread public suspicion of the process, and to promote it as an 'atoms for peace' technology untainted by the increasingly poor image of the main civil and military nuclear programmes. It is no exaggeration that the major pressure for the irradiation of food has come from the atomic energy lobby rather than the food industry as a whole. The process has been allowed in Britain since January 1991, subject to some rather unsatisfactory conditions regarding the labelling of irradiated food.

Despite many years of testing – blighted in the United States by major scandals over falsified results – there are still certain doubts about the safety of irradiated food. A study in which malnourished Indian children were fed on irradiated wheat showed consequent chromosomal abnormalities in blood cells, an effect subsequently replicated in studies with monkeys, rats and hamsters. Substantial vitamin losses occur at the recommended radiation dose levels (often around 25 per cent, but up to 90 per cent in the case of vitamin B_1), and the palatability of fatty foods is often affected by the development of curious 'off' flavours. Meat has been described as having a 'wet dog' smell, and milk products as tasting 'like burnt wool'.[51]

The real worries about food irradiation are not over the direct effects just mentioned, but over a number of more subtle microbiological hazards. At the sublethal dose levels employed, it is quite possible that radiation resistant bacteria will rapidly develop, just as antibiotic resistant strains have done. Repeated radiation under laboratory conditions has already produced resistant strains of salmonella bacteria. A more mundane source of danger is that, while irradiation might kill the bacteria that make stale food smell bad, and the micro-organisms which cause food poisoning, it cannot remove the toxins which they may already have produced. If the food being irradiated is fresh, this should not be a danger, but there have already been a number of cases of irradiation being used to 'freshen up' slightly decayed products (particularly shellfish), and many public health officials are concerned about the difficulty of detecting such abuses. At present, a relatively simple check on bacterial levels gives a fairly reliable index of food wholesomeness, whereas with irradiated foods the presence of dangerous toxins could only be established by exhaustive chemical analysis.

The moulds responsible for aflatoxin contamination of grain crops and other vegetables appear to be relatively resistant to radiation damage. There is evidence that irradiation increases the level of aflatoxins produced during storage in less than ideal conditions.[52] Another highly dangerous organism which can survive radiation at the level used for foodstuffs is *Clostridium botulinum*, the agent responsible for the serious type of food poisoning known as botulism. If the normal spoilage bacteria are killed off by irradiation, there is a risk that (under conditions where air is excluded) *Clostridium botulinum* could flourish to the extent that food became highly toxic without developing any warning smell.[53] Although relatively rare, cases of botulism are often fatal, and the food canning industry is subject to stringent safety precautions in order to avoid its occurrence. The risk of botulism is likely to be a major constraint on the use of irradiation for meat products. In the United States, the Department of Agriculture has indicated that normal vacuum packaging methods will not be permitted for irradiated meat until extensive safety studies have been carried out.[54]

Although irradiation may prove to be a valuable method of preserving certain raw ingredients against spoilage, the cards at present seem stacked against it gaining public acceptance for large scale use on ready-to-purchase foodstuffs. Even in the more limited context of the use of irradiation as a solution to the problem of bacterial contamination of poultry and meat products, it is not clear that investment in the large numbers of plants required would be justified. A Canadian study of ways to reduce salmonella counts in chickens rated the use of ionizing radiation as seventh out of eleven possible methods in terms of cost-effectiveness.[55]

Another new technology, whose health implications are even less clear than those of irradiation, is that of genetic engineering. The possible effects of the first widely used genetically engineered product, the bovine growth hormone, BST, have recently been discussed by the British Government's Veterinary Products Committee. According to Professor Richard Lacey, routine injections of dairy cattle with BST may distort bone growth in the cow and her calves, and impair kidney and liver function, as well as being uncomfortable for the cow. The human health hazards for consumers of BST milk include the unknown long-term effects of ingesting BST; the effects of other changes in the normal nutritional composition of the milk; the possibility of increasing the virulence of the mad cow disease BSE; and the unknown effects of mutant versions of the hormone which are inevitably produced in small quantities during its manufacture.[56]

The produce of factory farms, however attractive its price, is often accused of failing to measure up to the food we used to have in the good old days. This accusation can be levelled on grounds of taste, or lack of certain nutrient qualities, or because it is believed that intensively produced food contains poisonous chemicals or dangerous bacteria we should be better to avoid. How should we view the evidence? In terms of taste, it can hardly be denied that the best old-fashioned products had a richness and subtlety unmatched by the bland and insipid substitutes which now make up the bulk of the market. But at least in matters of taste, everyone can be his own judge, and while we may not agree with the person who can't tell a free range egg from a battery one, or who won't pay the difference even if he can, we cannot prove him wrong in such a personal verdict. On the more scientific level of nutritional adequacy and the possibility of harmful residues, we can look at the available evidence, but it may not be 100 per cent conclusive. Few nutritional defects have been pinned definitely on factory farm produce, but nutrition is not a very precise science, and ideas of the essential components of a healthy diet are subject to change (see chapter 4). The question of harmful chemical residues is also vexed: even though their presence is certain, their long and short-term effects on human health are the subjects of controversy. The bacterial residues of the factory farm, on the other hand, carry their own individual stamp, and indubitably give increasing cause for concern.

In medieval England, a jury had the choice of four verdicts: guilty, not guilty, *ignoramus* (we do not know), or *ignorabimus* (we shall not know).[57] The frustrating fact about toxicological testing of substances for their effects on health is that *ignorabimus* is often the only practical conclusion. To reach a definitive verdict on the long-term carcinogenicity, teratogenicity and mutagenicity of a suspected toxin would require controlling the diets of 100,000 or so people for twenty or thirty years, and following their health histories for the rest of their lives. For most substances, we shall never know whether they are completely safe; and the knowledge we do gain of their long-term effects is many years in the gathering. What, then, should we do when suspicions are aroused? There is no point in waiting for proof positive and incontrovertible, because that sort of proof may never be available. When H. J. Heinz decided to ban use of the 'dirty dozen' pesticides on crops for infant foods, they adopted a more cautious approach: 'A lot of these chemicals are going to be proved safe, but we don't know which ones.'[58] Taking account of reasonable probabilities, rather than insisting on absolute certainty, is a strategy which is both rational and prudent. Or as they say in my part of the world, 'Gang warily'.

Part III
MORAL CHOICES

7
Why Care About Animals?

At fiesta time in many a Spanish or Latin American village, a gory spectacle is enacted. Live chickens or geese are tethered to the top of a pole while the local braves take turns at hurling arrows or stones, or try to seize and pull off the birds' heads from horseback. How the little children clap their hands, and what pious tears of joy the mothers weep, to see the holy festivities!

In the still hours of darkness, many a sleepy English village is the setting for a pageant no less bloody. Long vehicles draw up beside vast dimly lit sheds, where teams of boilersuited workers are stuffing crates with chickens in their thousands and their tens of thousands. Many of the birds suffer broken bones or dislocated hips, and if the journey to slaughter is long, more will die from cold or thirst on the road. At the abattoir, the birds are shackled by the feet to an overhead conveyor which takes them, hanging upside down, to be stunned, bled and scalded. In theory, they lose consciousness in the electric stunning bath, but in practice many birds go to the knife inadequately stunned, and the unlucky few who escape having their throats cut are boiled alive in the scalding tank. This carnage, though, is simply routine: no laughing children or wailing mums witness, no symbolic prefigurement of calvary sanctifies, the broiler hen's mere workaday suffering.

What is the essential difference between these examples? Most people would dispute that they were in any way morally equivalent. (Let us assume that in accordance with social convention, the revellers actually do enjoy the slaughter while the chicken cullers are as grave as undertakers – an oversimplification perhaps, but necessary to clarify the argument.) In the first case, wanton and gratuitous cruelty is responsible for the birds' ordeal; in the second, their torment is ordained by the necessities of profit and expedience, and within the limitations of these parameters a degree of effort is expended to minimize the pain they feel. The gallants of the fiesta are moral agents whose conduct appears both repugnant and blameworthy by enlightened standards, whereas the impassive operators on the slaughterhouse line are hardly responsible for the suffering of the broiler birds under a system as impersonal as it is inexorable. Of course, the raucous Iberian heroes might claim that their exploits are just as necessary as the job of the piece-rate workers at the meat plant: their rituals, after all, have a longer cultural tradition than chicken and chips, and are doubtless of great emblematic importance. But such a plea mitigates rather than exonerates: cruelty is only marginally less cruel when it wears the respectable mantle of tradition. Aficionados of the dog-pit or the bull-ring can glorify their blood lust in aesthetic terms, and the promoters of rodeos who torture horses into a frenzy can do so in the name of Wild West tradition; but however socially acceptable these customs may be, they differ little in kind from the more random cruelty of naughty boys or brutal psychopaths.

From the birds' point of view, however, neither human justification of their torment, nor human enjoyment of it make any difference at all. What are we to make of it? Is the cruel relishing of their plight all we object to, or should we care equally about all animal suffering? The Christian church, when it has considered the issue, has tended to condemn cruelty to animals more on account of its corrupting effect on the perpetrator than because of the suffering of the victim; an attitude reflecting the general belief that animals do not have souls, and are therefore not morally considerable. Secular moralists, too, who have based their theories, not on religious dogma, but on the apparently more secure legalistic foundation of a social contract or basic human rights, have by this very token had little to say about the treatment of animals. And Kant, who tried to found morality on a base that was neither religious nor secular, adopted the conventional Christian position almost without change.

There is, indeed, something about acts of wanton cruelty which excites censure out of all proportion to the amount of suffering involved. In Ancient Rome the circuses were orgies of blood, yet Suetonius refers in disapproving terms to the Emperor Domitian's delight in killing flies.[1] And at the height of the carnage in Vietnam, sentimental Americans could still be shocked by their president picking up his pet dog by its ears. While these incidents would doubtless have passed unnoticed had not the observers feared that they might some day be the recipients of like treatment, the lack of extenuating circumstances gives them a curiously unpleasant character.

But the argument that cruelty to other species is wrong solely as a bad example which may encourage cruelty to humans collapses in on itself. If it really had no other significance, it could not possibly be a bad example. No-one would claim that the mineralogist who enjoys slicing sections of rock to display the pretty colours within should be regarded as a danger to society, who might soon turn to slicing up human beings. No objection to cruelty can hold water, if it is not accompanied by an objection to the resultant suffering on its own account.

The idea that other species have no moral value is an absurd exaggeration of the Christian doctrine that man, as the chosen vessel of the incarnation, is the centre and measure of the whole world. Recently, some theologians have reverted to the view that God cares about other parts of his creation as well as man, and have tried to develop the concept of stewardship to include duties towards the rest of nature. Many environmentalists, though, have rejected orthodox religion in favour of a kind of pantheism which finds spiritual values in nature. Their message, too, is that animals (and plants) have values in themselves, and not merely because of their use to mankind. Leaving God out of it, the utilitarian moralists had already proposed that moral status should not depend on 'reason', as philosophers like Hobbes and Locke had held, but on the capacity to experience pain and pleasure. On this basis, animals were obviously entitled to consideration. In Bentham's celebrated words, 'The question is not, Can they *reason*? nor Can they *talk*? but Can they *suffer*?'[2] In the past twenty years there has been an upsurge of philosophical interest in the issue of animal welfare, and following the publication of Peter Singer's *Animal Liberation*,[3] a number of other substantial books on the subject have been produced.[4–6]

Moral or religious systems can only justify conduct for those who believe in them. The question of why we should care about animal suffering is no more capable of a final answer than that of why we should care about human suffering: to the determined egoist no proof will be convincing, while for the good Samaritan none is necessary. But the fact remains that we do, on the whole, care. Most people would prefer a world in which animals were kindly treated and happy to one in which cruelty and suffering were the norm; and a moral ideal is simply the expression of a vision of such a better world.

How much we care depends on social as well as personal factors. What counts as acceptable conduct varies enormously from culture to culture – even within a given society, perceptions are likely to vary between country dwellers and those who live in the city. The farmer's relationship with his animals is primarily economic, and while understanding their basic needs is essential for the success of his enterprise, compassion will inevitably be tempered by the profit motive. The town person's experience of other species is more emotional; as he values the companionship of pets, or the pastoral vision, relatively untrammelled by economic constraints, and has a much wider choice between involvement and indifference. Time, as well as place, makes a difference: research in the United States showed that there was apparently much less interest in animal welfare during wartime than in periods of peace and prosperity.[7] For almost everyone, though, there is more to relations with other species than simply economic or emotional exploitation; just as there is more to our relations with other people. We do have some sympathy with other creatures in much the same way as we sympathize with other humans, and the 'species barrier' is not an absolute divide.

It may still be objected, by those of a practical bent, that it is unnecessary and silly to be concerned about suffering that is not the result of deliberate cruelty. If we are really going to get as worked up about the inevitable discomforts of broiler chicks in even the most efficient slaughterhouse, as we do over the sadistic revels of a lot of appalling peasants, where will it all end? Should we therefore weep for every sparrow the cat catches, and grudge the lion his every meal? Nature is cruel, and there's an end to it. Such objections completely miss the point. The lion is not designed, created or operated to fulfil human purposes: the slaughterhouse is. For his savagery nobody is responsible: for the foreseeable results of human actions, humans are to blame. He has to kill to eat, and cannot do so painlessly: we know how to reduce the broiler chicks' suffering, but choose not to. As for the jolly peasants, they are an easy target on which to fix the blame their whole society should rightly share: the absence of such blatantly guilty parties in our own milieu leads us too easily to the presumption that our more dispassionate system of bloodshed is entirely innocent.

If we – individually or collectively – are responsible for what we do, and make moral choices depending on the likely outcome of our actions for better or worse, then if any of these actions are the cause of suffering we cannot remain unconcerned. The fate of the production-line broiler fowl is unaffected by whether the person who pushes the button does it with sadistic delight, with bored indifference or with anguished regret. But this is not the whole story – that bird's suffering is still the responsibility of those who collude in it: the farmers, the food processors and the consumers.

I have used the chicken slaughterhouse merely as an example of how an impersonal system can cause more suffering than many individual acts of cruelty. Cruel people may be easier to recognize than cruel systems; but the majority of farm animals suffer far more from the latter. Before the welfare implications of modern intensive husbandry systems can be judged, though, we should consider more closely just how much we can tell about the ways in which animals experience pain, deprivation or other forms of suffering. Only on the basis of a thorough understanding of the welfare needs of different species – of what things are important for those particular animals – can different husbandry methods be evaluated and improved.

8

The Assessment of Animal Welfare

What, and how, do other animals feel? Just how different they are from human beings is debatable. According to one view, which was first proposed by the pioneering scientists of the post-Renaissance period, animals were soulless automata, as different from people as the clockwork canaries so popular at the time. This idea, which still has a wide currency, can partly be explained by the curious social conditions prevailing at the time of its original exposition. The foundations of modern science were laid under the shadow of the inquisition: in 1600 Giordano Bruno was burnt at the stake for denying the earth was at the centre of the universe, and ten years later Galileo narrowly avoided the same fate. The Church was certainly no friend of animals; many of the heresies so vigorously put down by the inquisition were connected with animal cults, or with the belief that other creatures had souls just like humans. In such a climate, the anthropocentric doctrine of the dignity and uniqueness of the human soul could be questioned only at the gravest risk.

One of the most important figures in determining the course taken by the new experimental sciences was the French philosopher René Descartes. He was well aware of the anatomical similarities between human beings and other animals, and the mechanistic view of the universe he proposed could have implied that men, too, were nothing but machines. However, he saved his scientific theories, his religious faith and his skin by re-emphasizing the importance of man's immaterial and immortal soul. The existence of the human soul was intuitively self-evident to Descartes, while the lack of rationality and linguistic ability shown by other animals was proof for him that they had no souls at all. 'I concluded that I was a substance whose whole essence or nature consists only in thinking, and which, that it may exist has no need of place, nor is dependent on any material thing: so that "I", that is to say, the mind by which I am, is wholly distinct from the body.'[1]

Such extreme dualism, in which body and soul occupy two entirely separate worlds, led to the peculiar conclusion that if animals do not have immortal souls they cannot have any sort of consciousness, or even be aware of pain. This was highly agreeable to the seventeenth-century scientists who had taken up the practice of experimenting on living animals: why should they have a conscience about horrible cruelties inflicted in the name of scientific curiosity, when they could believe instead in a fashionable theory which denied that animals could suffer at all? Living organisms were like clocks; 'the cries they emitted when struck were only the noise of a little spring that had been touched ... the whole body was without feeling'.[2]

Not all scientists of the time believed in this doctrine, but its sheer convenience ensured it had a long innings, and traces of it are still to be found in the beliefs of scientists and lay people alike. However, the most extensive research in physiology, neurology and anaesthetics has so far failed to reveal any qualitative difference in the reactions of humans and other animals to 'painful stimuli'. A well-established modern criterion is that 'Pain in animals is manifested by abnormal behaviour which can be alleviated by analgesic procedures which relieve pain in humans', while the journal *Laboratory Animals* claims quite simply that 'it should be assumed that if a procedure is likely to cause pain in man, it will produce a similar degree of pain in animals'.[3] An independent report on field sports sponsored by the Royal Society for the Prevention of Cruelty to Animals (RSPCA) recommended the application of this principle to fish as well as warm-blooded creatures; and if pain is regarded as a bodily rather than an intellectual condition, there is nothing particularly unscientific in extending the use of the term to lobsters, or even to the poor beetle that 'in corporeal sufferance finds a pang as great as when a giant dies'.[4,5] Few people are likely to attach as much importance to the suffering of fish as to that of pigs or cows, but it seems unduly dismissive to deny – as many anglers do – that they are capable of feeling pain at all. The apparently less complex minds of 'lower' animals incline us to regard their lives as of little value; and we may judge, from their slowness in learning to avoid unpleasant stimuli, that the feelings of fear and frustration which compound the effects of pain and injury in the brainier species are also relatively undeveloped. Relatively is the important word, though: there is no obvious cut-off point below which pain inflicted on other organisms becomes morally negligible.

It may be objected that such lack of restraint is an example of the 'pathetic fallacy' almost as extravagant as the attribution of feelings to a book or a stone; and that we can have no idea that other creatures feel pain in anything like the way we do. But neither, as Wittgenstein points out, can we be sure that other *people* feel pain as we do, just because they use the same word.[6] We are not usually troubled by sceptical doubts as to whether or not a dumb person, or a child who cannot yet talk, can actually feel pain – so why doubt that a calf or a chicken can? Nor can it be assumed, simply because pigs do not use concepts such as frustration or boredom, that they are incapable of being frustrated or bored, and suffering as a result. If the normal lives of other animals have any value at all, then manifest damage to these lives by injury or by deprivation must be accounted harmful, whatever our doubts about their related mental processes.

Because they cannot talk, we have an immediate problem assessing pain and suffering in other animals. It is necessary to rely on physical signs, rather than direct communication. But just as a doctor assumes that a severed finger or a high fever is uncomfortable, however stoical the protestations of his patient; so obvious damage or sickness can sensibly be connected with suffering in other species. (The connection is reasonable, but not necessarily infallible. There are many cases on record of horrific injuries where the patient apparently felt little pain, and conversely, persistent complaints of pain where no tissue damage can be found.) Obvious wounds, broken bones, blisters, and other such injuries present few problems of recognition; though their severity from the animal's point of view may perhaps better be judged with the aid of additional clues, such as whether it flinches or cries out, how its posture, breathing and heart rate are affected, and so on.[7] With many illnesses, abnormal behaviour or impaired growth are the first symptoms, and precise diagnosis depends on pathological tests. In almost every case, the starting point for recognizing pain or sickness is a departure from normal appearance or behaviour; and while anyone could probably spot that there was something far wrong with a beast on its knees and groaning, the more experienced stockman will spot signs of trouble much earlier.

Although the behaviour of many other species has much in common with normal human reactions to pain, hunger, thirst or extremes of temperature, there are inevitably some differences which are only revealed by experience, such as the particular vulnerability of pigs to strong sunshine, or the lack of sensitivity of a horse's mane. The more subtle behavioural cues differ between species, too: we tend to make a lot of use of facial expression for interpreting what animals are feeling, and while this often stands us in good stead, it is not always reliable, as many victims of attack by bears have found to their cost. With some species, very little is known about normal and abnormal behaviour patterns. It is only within the past few years, for example, that the reactions of fish to stresses or painful stimuli have been studied by filming in aquaria; previous knowledge having been restricted to observations of hooked or netted fish when they finally broke surface.[8]

A hundred years ago, when every farmyard was full of animals doing more or less their own thing, it would hardly have occurred to their keepers that there was any doubt over what constituted natural behaviour for cows, sheep or horses. Even then, there may have been a few ignorant boys in urban dairies or stables who had little idea of how the animals they were tending would behave in meadows or on the open range. But now, a whole generation of farm workers has grown up for whom the natural environment of laying hens is the battery, of breeding sows the stall, and of cattle the covered byre. It must be questioned how this will affect the ability of future stockmen to diagnose welfare problems. The behavioural repertoire of animals under conditions of close confinement is often so restricted that even inability to walk may not be apparent at the rather cursory inspection of stock which is the norm.

It is immediately obvious that intensively kept livestock do not behave in the same ways as those on free range, but there is considerable debate over how much this matters. Perhaps the battery farmer can argue that since his stock have grown up not knowing any other surroundings, they are not being deprived just because they have not adopted some of the habits of outdoor animals. There is no analogy with imprisoning a human being who has been used to freedom, or even with the incarceration of wild animals in zoos, since for what is unknown there can presumably be no desire. Against this argument, though, it can be urged that it would be counted a greater misfortune for a person to be brought up in a prison from which he was never released, and in which many of the normal social activities of our species could never be realized, than to lose an arm or leg, but otherwise live a free and full life. We should probably accept this, even if it were certain the prisoner never knew what he was missing. On such a view, it is reasonable to be concerned for crated veal calves, even if they were in perfect health; for they live a life as deprived of all social stimuli as that of a human infant kept in perpetual solitary confinement.

The 'behaviourist' school of animal psychologists – who based their studies very much on observations of creatures kept in cages – tended to assume that life was largely a matter of learning to avoid unpleasant stimuli. According to this criterion, life in the battery house with plenty of food, water and warmth could not be that bad. Since the war, however, much more emphasis has been given to the work of ethologists who have studied wild animals as far as possible in the natural environment that evolution has adapted them to. These workers revived the idea of instinct, and demonstrated the existence in many species of quite complex behaviour patterns which do not need to be learnt, but are apparently 'programmed' in the animal from birth, waiting only for the right stimulus to release them.[9] Nest building by birds is a typical instinctive behaviour pattern of this sort, which wild birds perform at the appropriate time without needing to be taught. When battery hens go through the movements of nest making or dust bathing, though they have no straw or earth, but only the wire mesh of their cages, it can be claimed that their natural drives are being frustrated, and that they must therefore be suffering. The Brambell Committee on intensive husbandry methods took this view. 'The degree to which the behavioural urges of the animal are frustrated under the particular conditions of the confinement must be a major consideration in determining its acceptability or otherwise.'[10] More recently, research on the wild ancestors of farm livestock, and on feral populations of domestic strains, has helped to establish what their natural behaviour patterns are, and this approach has thrown light on a number of husbandry problems.[11,12]

Welcome as this trend is, as a pointer towards possible causes of distress, it cannot necessarily be assumed that all unnatural behaviour necessarily involves suffering. If natural escape behaviour is not performed, simply because there are no predators around, it need not follow that frustration of this particular drive is bad for the animal. The highly unnatural life of human beings, while no doubt frustrating some of our basic urges, does have its compensations; and so, perhaps, does that of the captive animal. Gerald Durrell records how difficult it was trying to release some parrots he was not allowed to export from Paraguay: they refused to return to the wild, some of them eating through wood and wire to get back into their cages.[13] There are also practical limits beyond which no farmer could afford to indulge his livestock's natural life style. Hens that go broody and cows that go dry may be acting delightfully naturally, but they are not earning their keep.

As well as being unable to perform some of their natural behaviour patterns, it is often observed that animals in captivity go through certain recognizable sequences of actions which are not seen in the wild. These new habits can often be related to boredom or some specific deprivation, though they are not necessarily unpleasant in themselves. Chimpanzees in zoos, for example, apparently enjoy throwing faeces at or urinating on the ranks of watching humans.[14] More typically though, many zoo animals pace endlessly back and forth in an abnormal behaviour pattern often referred to as a stereotypie. Other stereotypies exhibited by confined animals are endless gnawing at their cage bars by battery pigs, and suckling of each other by bucket-reared calves. Despite the lack of satisfaction, such apparently pointless actions may be carried on indefinitely, and have generally been regarded as signs of distress. Autistic children display similar stereotypies, and 'abnormal' behaviour in general is often, though not infallibly, linked with unhappiness or mental unbalance. (Hediger, in his pioneering work on zoo animals, reports that he had occasion to observe a man who insisted on urinating into the mouth of a moose, and another who distributed religious pamphlets to the snakes – actions at least as difficult to explain as those of the average dumb animal.)[15]

Animals in the boring environment of an intensive livestock house often appear unnaturally inactive, or 'apathetic'. By studying the responsiveness of sows in different types of housing to novel stimuli, such as the pouring of a small amount of water on their backs, Donald Broom has demonstrated that confined sows are 'less responsive to events in the world around them' than group-housed animals.[16] Impaired ability to carry out normal behaviour is another indicator of poor welfare. Studies of cattle on slatted floors, for example, have shown that lying down is often preceded by many unsuccessful attempts; the whole procedure lasting up to 20 minutes.[17]

Behavioural differences between captive and wild animals are an invaluable starting point for investigating possible welfare problems, but unambiguous interpretation is not always possible. One source of corroboration which seems hopeful would be the development of physiological methods of measuring suffering. Since animal husbandry does not on the whole involve the deliberate infliction of pain – the mechanism of which is, in any case, very imperfectly understood – most early attempts to relate welfare to physiological changes have concentrated on the effects observed in animals that are put into situations assumed to be unpleasant rather than painful. It is known that when an animal is suddenly confronted with danger, it reacts initially with the 'adrenalin response': secretion of adrenocortophic hormone (ACTH) stimulates the adrenal glands, heart rate often increases, and breathing becomes deeper as the body prepares for violent muscular effort. If the danger continues, without the need for fight or flight, the animal remains tense, and in a constant state of readiness. Eventually, exhaustion sets in, characterized by such effects as a decline in growth rate, decreased sexual activity, gastric ulcers and pathological enlargement of the adrenal glands. This whole sequence of events was described by Selye as the 'general adaptation syndrome', and he suggested that reactions to 'stressors' such as heat, cold, pain and bacterial infection, all followed a similar pattern.[18] Although it has since been shown that the detailed physiological response is not the same in all these cases, the concept of a measurable index of 'stress' which could enable objective differentiation between the welfare of animals in different situations is obviously attractive.

The short-term reaction to unpleasant stimulus, in which the animal prepares for flight, defence or hiding, has been studied in a number of species using measurements of such variables as heart rate or blood cortisol levels. The former has the advantage that, by use of attached miniature radio transmitters, observations can be made without the need to handle the animal, an action which may itself cause stress. It has been observed that sheep show a large heart-rate response to the approach of a dog, and that the use of an automatic gathering machine to collect broiler chickens causes a much briefer disturbance to the normal heart rhythm than collecting and packing the birds by hand.[19,20] Many observers have tried to correlate changes in blood or brain chemistry (measured by analysis of blood samples or during post-mortem examination) with the degree of stress caused by various handling operations.[21] One problem with measurements of this sort is whether or not all such 'stress' is a bad thing. Adrenal cortical activity is equally stimulated during courtship and mating, which activities are nevertheless eagerly pursued by most creatures. There is also evidence that a certain amount of stimulus increases disease resistance, and that left to themselves animals actually seek out some types of stressful situations.[22]

A number of attempts have been made to detect physiological abnormalities in animals suffering from long-term discomfort or deprivation. Statistically, the incidence of disease under particular regimes can be used as an indicator of welfare (see page 126), but it has proved more difficult to devise tests which would show the general level of well-being of the individual animal. It has been shown that under prolonged adverse conditions, bursts of exaggerated adrenal cortical activity often persist, with consequences measurable by what is termed the 'ACTH challenge' technique. The functioning of the immune system has also been correlated with long-term stress; and measurements of sows' antibody production following injection of sheep red blood cells, or other similar reactions to foreign substances, have been proposed as indicators of welfare.[23]

Recently, as the physiological basis of human pain has become clearer, evidence has been found which suggests that the gap between physical pain and stresses such as boredom is not as great as might have been supposed. Severe pain can apparently be controlled by secretion from the hypothalamus of a morphine-like neurohormone called β-endorphin. In the case of battlefield injuries, this analgesic mechanism can be so effective that pain sensations temporarily disappear altogether. This sort of analgesia has been investigated using the observed fact that it can be blocked or reversed using the drug naloxone, which is used to treat heroin addiction.[24] When pigs showing the stereotyped behaviour associated with close confinement were dosed with naloxone, it was found that the stereotypies were suppressed, and the pigs reverted to the 'normal' behaviour (i.e. aggressive displays and efforts to escape) that they had shown when first put into the stalls.[25] Direct measurements of β-endorphin in blood samples from lambs has also shown that levels of this hormone are related to 'stress'.[26]

Despite the advances which have been made in the study of our sensations, knowledge of human physiology is a long way from providing a quantifiable biochemical index of happiness or misery, and until it does we can hardly be confident of the application of this sort of method to other species. As with abnormal behaviour, physiological abnormalities are a valuable pointer to areas of concern, but not an infallible sign of suffering.

An interesting alternative approach to the welfare assessment of different husbandry systems has been developed at Oxford by Marion Dawkins. Instead of looking for measurable signs of suffering in animals kept under a particular regime, she sets up a situation in which the animals can choose themselves between different environments. For example, hens were given the choice between a battery cage and an outside run, and the results showed that birds which had been used to living outside invariably preferred the run, while those accustomed to cages tended to choose the cage at first, but changed their preference after a few minutes' experience of the run.[27] The strength of the hens' preferences can be gauged by putting food in the less favoured environment, or making them run the gauntlet of stimuli they would normally avoid before they can get into the more favoured area. Similarly, pigs which are allowed to control the level of heat and light in their houses by snout-operated switches soon learn to make their own choices of these environmental factors.

Such experiments can yield very useful results, as long as care is taken with their interpretation. Choices can be affected by the animals' past experience, and the strength of preferences may vary according to the time of day, degree of hunger, or reproductive state of the subjects. While this sort of investigation is probably the best way we have of finding out what animals *like*, it does not always follow that what they choose is necessarily good for them. Like human beings, other species often show strong preferences for sugar, saccharine and other addictive substances; and it is also pretty clear that many animals have a strong aversion to being treated by the vet. However, for investigating the physical environment most agreeable from the animals' point of view, this sort of technique is difficult to fault. It may be argued, of course, that merely because one environment is favoured, however strongly, this does not prove that the other will necessarily cause suffering. Merely because I should prefer a Georgian villa, it does not prove that I should be hurt by having to live in a council flat – indeed, I might soon become so used to it that I no longer wanted to change. To such a niggardly attitude there is no straightforward reply; but if a particular environment inspires such determined rejection that the animal will sooner face pain or hunger, it is surely more reasonable to assume that there is something *wrong* with that environment, rather than merely something not quite right.

In discussing how the welfare of animals in different circumstances can be judged, I have emphasized the difficulties of the various methods available. In many actual cases there is no problem: maltreatment very often results in physical damage, gross behavioural abnormality or decreased life span, and methods of husbandry which have such outcomes can be condemned on these grounds alone. Regimes which would have similar effects if the animals were not tranquillized or regularly dosed with antibiotics must also give grounds for concern. It is only where quantifiable harm is not apparent that scientific proof of suffering becomes so difficult. In these cases, as in doubtful cases of human deprivation, we should simply judge as well as we can pending further research, and be generous in giving the benefit of the doubt.

The welfare of transgenic farm animals may be somewhat more difficult to assess than that of the traditional breeds of livestock. If cruelty is defined as the infliction of pain, such a criterion could readily be applied to most genetically engineered animals. There might be borderline cases, but no more so than already exist regarding the degree of sentience of fish or molluscs. But the relation of behavioural needs of captive animals to the natural behaviour of their wild cousins would clearly be more difficult. Although many existing breeds of farm livestock are apparently quite different from the wild strains they are descended from, such ethological studies as have been done show quite close similarities in behaviour between the wild type and domestic animals put in approximately natural surroundings. With new transgenic species, such behavioural links would be much more difficult to establish. The problem would be more like that which already confronts us with conventionally bred turkeys having so much breast muscling they are unable to mate, or Belgian Blue cattle whose leg muscle conformation can make normal delivery of the calf impossible.

With such animals, that are frankly non-viable in the wild, the relation of welfare to natural behaviour cannot be pressed too far. Welfare legislation which refers to 'natural' needs is in danger of becoming meaningless. But rather than despair, there are several alternative strategies to fall back on for establishing a basis for humane treatment of transgenic animals. Zoo workers frequently have to make judgements, based on general experience, for species they have little or no knowledge of in the wild. However much detailed behaviour patterns vary, such basic needs as freedom from injury, disease, malnutrition, fear, stress, thermal or physical discomfort, and severe restrictions on behaviour, can still be assumed to apply see (see page 205). The precise interpretation of these 'freedoms' is open to question for any species, genetically engineered or not; but it might reasonably be assumed, for example, that an animal with the ability to walk or fly that is kept in conditions where it cannot do so is having its behaviour severely restricted.

The other methods of welfare assessment discussed above are as applicable to transgenic animals as to any others. Tests for physiological abnormalities have the same sort of indicative value that something may be wrong; but the further the organization of the animal differs from that of a human being, the less probability of translating physiological measurements into feelings without falling back on the check of behavioural symptoms. Choice chamber experiments, in which different factors are tested as positive or negative reinforcers in learning experiments, could establish clear preferences as regards simple environmental factors.

A further question which can be raised over at least the finer points at issue, is how much they are affected by differences between individual animals. In the case of human beings, personal preferences might well be considered important: for example, some people have a particular horror of drowning, others of being burnt alive, and so on. In the human context, there is no doubt that the cruelty of a particular method of killing, torturing or punishing is partly dependent on the subjective preferences of the individual victim. Similarly, there are enormous differences in how different people react to solitude, institutionalization, feeding routines, etc. – to the extent that kindness in one case might be counted as cruelty in another. How much is this relevant in the case of other species? Does one rule invariably apply to all cows or all pigs, or should they too be treated with consideration for their individual idiosyncracies? In the context of factory farming, such a question appears absurd. The whole idea of applying mass production methods to animal husbandry is that the beasts can be treated as interchangeable cogs in the machine. The mass of laboratory-based scientific knowledge about animals reinforces this way of looking at them. Such knowledge is valuable because it is general in application to all members of a species: variations between individuals are as much a nuisance to the zoologist in search of valid generalizations as to the farmer who needs standardized pigs to fit standardized cages.

But it should not be forgotten that this is not the only way of looking at other species. It is one way of understanding and using animals, just as physiology and economics provide a framework for understanding and using people as if they were interchangeable. That it is not the only way was well known to all stockmen in the past, and is still essential knowledge for those dealing with the temperamental dairy cow. Identical treatment may not be the way to get the best out of different individual animals, nor need it be equally kind to them all. For the town dweller who has little contact with sheep or chickens, it is an understandable belief that they are all pretty much of a pattern, and all the more convenient because such apparent lack of individuality does not encourage putting a high value on their comfort or happiness. When it comes to the family dog or budgie, however, such feelings are put aside; and few pet owners would feel that Fido or Peter could adequately be replaced by another representative of the species. Perhaps we are too keen, in other cases, to conceal our ignorance of significant differences by simply denying that there is anything to know.

Regarding animals as interchangeable atomic entities also neglects the fact that, like us, they are partly defined by the society in which they live. An early weaned calf may differ substantially in behaviour and welfare requirements from one which has grown up without disturbance of the parent–offspring bond. Large groups of animals all of the same age do not occur in nature, and it has been suggested that they encourage squabbling. The presence of a dominant cock has been shown to inhibit expressions of aggression among all birds in a population of feral fowl, up to a distance of 6 metres; and Kiley-Worthington reports a reduction in agonistic behaviour among horses which are kept in extended family groups.[28,29]

Because human beings have their own different ideas about what is pleasant or unpleasant, and their own particular hopes and fears, the welfare debate is bound to go on forever. The only fundamental reason for not hurting animals is that we ourselves should not wish to be so hurt: the golden rule of 'do as you would be done by', on which all morality ultimately rests. There is no way in which such a doctrine can be made precise, for not only can we never be certain of the feelings of another creature, but we have different ideas about just how important these feelings are. However intelligently we try to interpret the wants and needs of other species, how can I, who find little wrong with the prospect of a life of confinement, agree with you, who would give all for freedom? In the last resort, there will always be a subjective element in our judgements, and we shall inevitably be a little like the old maid in the rhyme:

> Ten cocks for ten hens!
> Was the spinster's cry:
> I know what it's like
> To be passed by.

9
Welfare in Modern Farming

The claim is often made by exasperated farmers that the concern of animal welfarists is essentially unfounded. Hard-nosed self-interest, they say, is as effective a safeguard for farm livestock as the attention of any number of do-gooders. 'Attention to animal welfare and successful poultry farming go hand-in-hand', complacently burbles a publicity leaflet from the National Farmers' Union. On this presumption, Bernard Matthews' £350,000 pay cheque is as certain evidence of the care he gives his turkeys, as the worldly success of the protestant merchant classes used to be deemed a token of God's approval. There is, of course, a degree of truth in the farmers' claim – the most cursory glance at history will show that when agriculture has been unable to prosper, both the animals and their masters have often suffered horribly – but the connection between welfare and profits is by no means hard and fast.

On the whole, injuries, disease, malnutrition and poor housing conditions are likely to reduce the biological efficiency of the animal, and hence impair productivity measured in terms of food inputs and outputs. There are notable exceptions to this generalization: 'sweat-box' pigs put on weight rapidly in conditions few people would regard as humane; and the laying hen will keep up normal production even when so crippled by 'cage layer's fatigue' that she can barely stand. Even where biological productivity can be used as an indicator of welfare, it is not invariably to the farmer's advantage to optimize the performance of his animals at any cost. Many poultry farmers find it more economic to leave their flocks unattended over the weekend; and there are numerous chronic diseases of all types of livestock for which the loss of production is usually less than the cost of veterinary treatment. Reduced labour costs and generous tax allowances can make intensive livestock systems economically attractive even if feed conversion rates and mortality are worse than for other systems. It cannot, therefore, be assumed that financial prudence on the part of farmers will necessarily ensure the well-being of their beasts. In recognition of this, there has in recent years been an increase in the use of the methods described in the previous chapter in an effort to place welfare standards on a more objective footing.

While it may never be possible to define a precise point on the welfare continuum, such that any better state is deemed satisfactory, and any worse one unsatisfactory, it seems reasonable to seek some sort of datum to which assessments of well-being or deprivation can be referred. One possibility is to consider the state of the animal in the wild, as has often been done in research on natural behaviour patterns. As a reference level, though, 'the wild' is very unstable. In sunny weather our hens pour out of their shed in the morning and apparently prefer scratching for worms all day to staying inside; but when it rains the great outdoors loses its appeal, and they all rush back to corn and dry straw. Not that this voting with their feet means very much – it might be a different story with birds unused to human society, or if they were surrounded by genuine jungle rather than blasted heathland – but it does show up the difficulty of defining a unique 'natural' level of welfare. Most animals in the wild are cold or hungry at times, and many die from these causes; others meet a sticky end by accident or predation. The life span of wild animals is often less than that of domesticated ones, and there are no vets to look after them should they fall ill. All in all, nature is a chancy business; even if enough of each species usually win through to continue the propagation of their race.

It will be much more convenient, in the following discussion of the welfare implications of modern livestock methods, to refer whenever practical to accepted alternatives, which may be better or worse, than to look in vain for any absolute points of reference. On the whole, such alternatives will be the traditional ways of keeping livestock, assuming adequate feeding and reasonable standards of stockmanship. Sometimes there are novel methods of husbandry which have been proved viable, and offer improved welfare prospects. Only when such comparisons are impossible, as in the case of fish farming or deer farming, will reference to life in the wild be unavoidable.

Confinement of the livestock is an essential part of most husbandry regimes, but the degree of restriction imposed on the animals can be anything from minimal to extreme. Non-migratory animals do not necessarily stray far from home even in the wild, and if their enclosure offers food, water and shelter, confinement need not be a cause of hardship. The fences which keep livestock in also fulfil the useful function of keeping out predators and competitors for food resources. As confinement becomes closer, though, overcrowding begins to cause welfare problems, even to the extent of preventing normal bodily movements. When animals are kept inside, it may be difficult to keep their housing at a comfortable temperature; and the fumes from their accumulating excrement sometimes produces chronic poisoning symptoms such as tender or damaged hooves, swelling around the joints or haemophilia.[1]

Poultry husbandry systems vary enormously in the space allowed to the birds. Free range growers and layers exhibit similar foraging behaviour to the jungle fowl from which they are descended, spending much time walking about, pecking and scratching, even when ample food is provided.[2] Their foraging range is seldom much more than 50 metres; at low stocking densities the boundary fence is needed more to keep mink and foxes out than to keep the birds in. At the maximum recommended stocking level of 1000 per hectare for free range layers, there would doubtless be a tendency for the birds to spread out further, more for want of suitable forage than to get away from each other. Battery houses are sometimes defended as providing the birds with protection from climatic extremes; but if housing of a reasonable standard is provided, free range birds can shelter as much as they want and keep warm at night, and in fact they normally choose to go outside in dry weather, however cold. (This inevitably means they tend to eat more to keep themselves warm.)

Life in the broiler house is very crowded, and the environment is monotonously warm, dimly lit and devoid of stimulus. For the very young birds there is adequate space, but as they grow bigger their movement is restricted and they may have difficulty getting to feeders and drinkers.[3] Feather pecking and cannibalism may also be immediate results of overcrowding; while the 'controlled environment' encourages rapid growth, it may be very bad for welfare. Inadequate ventilation often results in gross overheating in the summer, with attendant mortality; and if the litter is too wet the hocks and breast (which take the weight of the sitting bird) can develop painful ulcers. 'Capped' litter, on which a hard wet skin has developed, may result from bad ventilation, water spills or diarrhoea caused by inappropriate diet or disease. The resulting burnt hocks are a serious welfare problem, which affects up to 20 per cent of broiler chickens.[4]

Environmental pressure in the battery house is even more severe, and while death comes early to the broiler chicks, laying birds have to endure their lot for a year or sometimes two. In a normal battery cage the hens cannot stretch their wings or stand up straight, and must spend their entire lives standing or lying on a wire mesh floor. The fine wire mesh used for the bottom of the cage was condemned by the Brambell Committee, but a study in which the hens were allowed to choose between a standard floor type and the alternative heavy gauge mesh they proposed showed that the latter was actually less comfortable.[5] That the choice made by the hens in this research was merely the lesser of two evils has been demonstrated by other experiments which make it clear that hens tend to avoid wire mesh altogether if given the option of anything better.[6,7] Research at Oxford has also shown that with unrestricted head-room, 25 per cent of a hen's head movements are above 40 cm, the cage height recommended by the EC.[8]

A striking feature of hens in batteries and broiler houses is their apathetic appearance and general lifelessness. Free range birds are always on the move, often squabbling and usually noisy by comparison. It has been observed that given the choice, hens prefer to be on their own or with one other hen, rather than in groups of four or five. In larger groups, they become more quarrelsome as they get more crowded, down to a space of about 800 sq cm per bird, below which they become suddenly quieter.[9] These results have implications for intensive systems where the birds are crowded, but not in cages, suggesting that fighting may be more of a problem at intermediate stocking densities than it is either in battery cages or on free range. The apathy of caged layers is further illustrated by the fact that if their feeding is deliberately interrupted they show far less sign of frustration or bad temper than birds which are free to move around. They act, in general, as if they are bored to death – a supposition corroborated by an experiment in which the hen's need for general stimulation was shown by teaching the birds to work for food by pecking at a disc. Even when given free access to food, the hens preferred to 'work' for some of their rations.[10]

As well as being cramped and uncomfortable, battery hens often suffer damage to their skin and feathers from the hard wires of the cage. The lack of space for exercise is believed to be responsible for 'cage layer fatigue', the progressive bone disintegration which means that at slaughter innumerable battery hens suffer painful fractures during transport. Fatty liver syndrome, which is a frequent cause of death among laying birds, may also be partly due to lack of exercise.[11] The close contact of the hens with each other makes it difficult for threatened birds to escape attacks by others, and serious injuries may occur which are not easily spotted in the crowded conditions of the cage. Unfortunately for the hen, it takes an extraordinary amount of stress to reduce their productivity. Even in severely cannibalized birds, with infected wounds right into the body cavity, a high proportion still lay eggs normally.[12]

It is still possible to find people who believe that battery chickens are quite happy, and that the cages are good for the birds; but reputable scientific opinion is now overwhelmingly against this. Not only do caged birds have more chance of contracting various uncomfortable or fatal diseases, but their individual and social behaviour is severely restricted. They cannot walk, fly, stretch, dust-bathe, make nests or forage; all they can do, in fact, is eat, sleep and lay eggs. Even the conditions allowed for egg laying are arranged solely for human convenience, so that according to the Nobel prizewinning ethologist Konrad Lorenz, laying becomes an ordeal: 'For the person who knows something about animals it is truly heart-rending to watch how a chicken tries again and again to crawl beneath her fellow cage-mates, to search there in vain for cover'.[13]

Recent research into alternative egg production systems has resulted in a number of experimental regimes designed to improve the birds' welfare without sacrificing too much of the efficiency of the battery method. These include percheries with mainly wire mesh flooring, deep litter houses and covered yards provided with straw.[14] In all these systems the birds can move around freely, and it may be presumed their welfare would be substantially better than in cages. So far, none of these alternatives is in widespread commercial use in the United Kingdom.

The close confinement of pregnant sows is an environmental restriction even more severe than that of the battery cage. Pigs in stalls cannot turn round, and their only choice is to stand up or lie down, usually on a hard floor with no bedding. Lameness and sores are common results, as is stereotyped behaviour such as bar chewing. The pig is a highly intelligent animal, quick to learn new tricks, and capable of being trained like a dog to round up cattle or game.[15] In the wild they spend up to six hours a day foraging and exploring their surroundings. Physiologically (and, it is said, gastronomically) they are the farm animals most like human beings. The behaviour of sows in dry and tether stalls has been the subject of extensive study, and a recent review of the scientific evidence concluded that close confinement was the cause of severe distress.[16] The way in which the animals adapt to life in the pens shows resemblances to the development in humans of chronic psychiatric disorders (see page 113).

Sows kept indoors in groups frequently fight, and this is one reason why stalls became popular with farmers. Considerable efforts have been devoted to ways of overcoming this problem, which is often connected with competition for food, by providing various novel feeding arrangements. In one particularly promising development, each sow is given an electronic tag which enables her to operate an automatic gate and gain access to a protected feeding area where she receives the appropriate rations. Even with this system, however, a sterile concrete environment appears to encourage fighting to relieve boredom, and better results are obtained if some straw is provided as a distraction.[17] Even in outdoor

A modern advance in pig management is the use of electronically operated feeders, which control the food dispensed to each animal and thus reduce the chance of fighting among loose-housed sows. (Author)

herds, sows will fight over food if it is just dumped in a heap at one point, but if it is spread out there will not normally be any problem. As Richard Guy of the Real Meat Company remarks, 'an aggressive sow will chase a timid sow for 11 yards and a timid sow will run for 12 yards'.[18]

However the sows are housed during pregnancy, most farmers use some sort of farrowing crate to restrain them while the young piglets are suckling, and thus reduce the chance of the mother crushing her infants. Given a free choice, most sows appear to prefer a dark corner with a nest of straw, and the primitive contraptions often used to pin them in a fixed position on a concrete floor must inevitably cause hardship.[19] Breeds vary substantially in their maternal competence, and until this factor is given consideration by breeders, some sort of restraint may be essential. However, there is also evidence that confinement in farrowing crates leads to more piglets being born dead, or dying immediately after birth, possibly due to prolonged labour resulting from a hormone imbalance, and that sows which have not been kept in stalls are much less clumsy as mothers, so the situation is not entirely clear.[20]

Growing pigs, especially those weaned early, are sometimes kept in battery-type wire cages for four to six weeks after weaning; and this system has been condemned by the Farm Animal Welfare Council as environmentally unsuitable.[21] The design of pig houses for indoor rearing varies considerably; from the unacceptable 'sweat-box' with its gross overcrowding, concrete floor and absence of bedding, to superior buildings with ample space and straw bedding. Outdoor rearing of growing pigs is much less common, though presumably more agreeable to the pigs if they are provided with comfortable shelters.

The appalling living conditions of crated veal calves have received such publicity that this method of rearing has been banned in the United Kingdom. It cannot be doubted that the animals suffer as a result of the degree of deprivation to which they are subjected. An American study of veal calves showed major changes in blood chemistry in crated calves compared with those kept in individual hutches with yards. The calves in close confinement required five times as much medication as those kept outside, and also exhibited considerably more lameness.[22] The banning of veal crates in Britain is a somewhat limited welfare victory, since many British calves are exported and end up in crates belonging to Dutch and French veal producers.

Cattle kept outside are subject to stresses from weather and ground conditions, which are particularly pronounced in the case of animals kept in feedlots without proper shelter. Indoor housing during the winter months may be in the animals' best interests, though the willingness of housed cattle to go out and stand in the rain suggests that some sort of sheltered yard might be preferable.[23] Slatted floors, which are provided to maintain hygiene with the minimum labour, are not necessarily comfortable for the animals: reluctance to lie on the slats can mean the cattle take up to 20 minutes attempting to lie down.[24] Nor are slats as healthy for the animals as traditional straw bedding. The increasingly common complaint of environmental mastitis is caused by organisms which get into the cow's teat canal when she lies down on a wet or dirty surface; and a study of 15,000 cattle in eighteen different farms has shown that the overall incidence of disease was almost twice as great among the beasts which were kept on slatted floors, compared with those in straw yards. Lameness was the complaint most significantly increased, but eye and skin diseases were also worse among the cattle on slats.[25] The importance of a comfortable bed for cattle is shown by the fact that when offered a choice of bedding materials in cubicles, they will choose a good mattress rather than warmth, even at air temperatures as low as −20° C.[26]

Dairy cattle are worked very hard to maximize milk production, so it cannot exactly be claimed that they do nothing all day; but the indoor environment is monotonous, and this can lead to over-reaction to strange people or unfamiliar noises.[27] It has long been known that changes in routine can affect milk yields, and some dairy farmers lay on piped music to encourage their cows; but high yields are not a total guarantee of welfare, and a poor physical environment is undoubtedly to blame for many cases of mastitis and lameness.

The social environment is also important to cattle, and no change is more difficult to adapt to than the weaning of the young calf when it is just a few days old. Deprived of their mother's udder, such calves lick their own coats, or suck at the ears, navel or penis of other calves. Introducing new cattle to the herd, or breaking it up, can substantially lower milk yields, and may thus be presumed to have an upsetting effect.[28] Again, though, it may be dangerous to correlate welfare too closely with milk yield or weight gain, and any departure from traditional practice should be subject to careful scientific assessment of its welfare implications.

The winter housing of sheep is one innovation which should, on the face of it, improve the animals' environment. Hill and upland sheep are often undernourished in winter and even when supplementary feeds are provided the flock may be inaccessible for long periods. Research at Edinburgh has shown that, even when given relatively generous space allowances, sheep under cover are much more aggressive than in the open; and work is continuing to produce a design of housing which minimizes the stress on the animals.[29] Similar research has been recommended by the Farm Animal Welfare Council into the design of appropriate housing for other species such as mink and rabbits.[30] It must be said that such research, and the basic behavioural studies on which it relies, come rather late for such species, which have been farmed intensively for many years. However, perhaps it is too much to expect that animal welfare will be considered before new developments, rather than after.

At least deer farmers have tried to address welfare problems at the inception of their new form of husbandry. Already there are welfare codes, and parliamentary orders governing the conditions under which deer are kept. To date, transport and slaughter, which will be discussed below, appear to be the main problems.

In the equally new industry of fish farming, there is no official acknowledgement that any welfare problems exist. Many people deny that fish can feel pain at all, and if it is not always easy to sympathize with birds or mammals, how much more of a gap there appears to be between ourselves and the silent, cold-blooded fish. But recently published research by scientists at Utrecht University concludes that fish are capable of experiencing both pain and fear; and a German angling society was recently fined DM1200 for using live fish as bait.[31,32] In Britain, this practice is also prohibited; but if we do admit that fish can suffer, then surely many aspects of fish farming must give cause for concern. Unlike other farm animals, the salmon is a migratory species, and fish swim many thousands of miles between leaving their home river and returning there to breed. This is yet another point of difference between us and them, but it can be imagined that thwarting the migratory instinct may cause the fish to suffer. The dorsal fins of farmed fish are often completely eroded, which is normally attributed to damage by fish lice or abrasion as the fish rub against the cage netting as a response to irritation caused by lice. In the absence of detailed studies of behaviour, there remains the possibility that such damage is the consequence of stereotypies performed as a reaction to confinement.

However suitable the environment, so-called production stresses can still affect livestock adversely. The improvements in output achieved by modern farmers have been largely due to the development of new high yielding breeds and specially formulated feeding stuffs. By selectively breeding for rapid growth rate, or higher output of eggs or milk, other characteristics essential for general health and hardiness may be lost. With overbreeding, the general physique of the animal may not be able to keep up with the extremely rapid growth rates that are achieved. Broiler chickens' lungs cannot always supply enough oxygen to the bloodstream, while with other species the ability of the gut to absorb enough food is a limiting factor. There is evidence, too, that the high energy concentrated feeds used to boost production can result in serious health problems.

Both genetic and dietary factors have been implicated as possible causes of disease in laying hens.[33] Selective breeding for increased egg yields tends to produce more aggressive birds, and may contribute to the high proportion of tumours and fatty livers in end-of-lay hens. Layers are normally killed at two years old or less – just about a third of their natural life expectancy. A diet consisting entirely of mash often causes severe ulcers, and experiments at the Poultry Research Centre at Roslin have indicated that these cause the birds considerable pain, and are thus a very real welfare problem.[34] Diets including poultry offal are almost certainly to blame for the prevalence of salmonella in laying and growing hens. As well as being an obvious human health hazard, the disease presumably has a debilitating effect on the birds themselves.

The battery hen requires a certain robustness to cope with cage life for a year or more, and given a salubrious environment ex-cage birds often live to a ripe old age. With broiler chickens, selective breeding for rapid weight gain has produced birds which would hardly be viable at all outside the carefully controlled environment in which they are reared. Mortality in laying hens during their productive lives runs at around 6 per cent over a period of fifty-two weeks or more; many broiler flocks have higher death rates in the space of as many days. Up to 25 per cent of broilers are lame because of selection for rapid growth, and possibly due to the use of rapeseed meal in feeds.[35] Another genetic peculiarity of broiler birds is the phenomenon known as stunting syndrome. A percentage of the birds grow only about half as fast as the others, show poor feather growth, and eventually die. These birds should be killed as soon as the symptoms show: one way of doing this, it has been claimed, is to raise the water dispensers out of reach of the runts, so that they simply die of thirst.[36] The massive growth potential of broiler birds creates difficulties with the breeding hens that lay the eggs from which the broiler chicks are hatched. To avoid cardiac stress from excessive fatness, and haemorrhagic diseases associated with over-feeding, these breeding birds are kept on short rations, 'in a permanent state of hunger'.[37] To compensate, they tend to drink excessively, unless their water supply is also restricted so that they are permanently thirsty as well.[38]

Turkeys have also been the subject of intense genetic development, with the same potential effects of skeletal problems and susceptibility to disease; and these may be paralleled to a greater or lesser extent with other types of poultry and game birds. Among traditional forms of poultry husbandry, *foie gras* production must take all the prizes for cruelty. The diseased livers of force-fed geese are not an incidental mishap, but the deliberate objective of the *gavage* process. Konrad Lorenz, the world authority on goose behaviour, was dismayed by 'the tortures to which these highly organised animals are subjected'.[39]

The high milk yields achieved by modern dairy farmers have inevitably put extra stress on the cattle. Most dairy cows are now sent for slaughter before they are six or seven years old, although their natural life span is nearer twenty years. Replacement of grass and hay by silage and concentrates such as soya bean, fish meal or dried poultry manure encourages milk output, but places a strain on the cow's gut which is designed to cope with slowly fermenting forages. The quickly fermenting concentrates can cause rumen acidosis and consequent lameness. Professor John Webster compares the modern dairy cow to a 'highly tuned racing car designed to run as fast as possible on very high grade

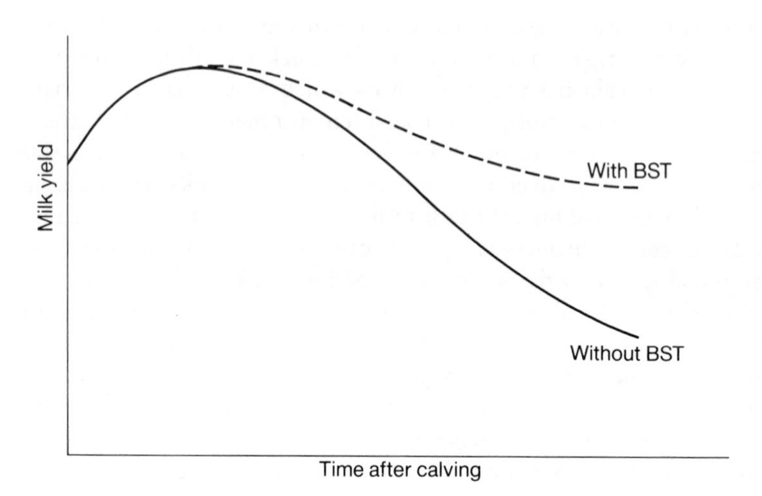

Figure 3 Effect of regular injections of the genetically engineered hormone BST on the milk yield of the dairy cow.

fuel', and comments that while the results can be very impressive, they may also be catastrophic.[40] The proposed use of bovine growth hormone (BST), which was introduced in controversial secret trials in Britain, and is already in widespread use in America, has been criticized by many veterinarians, including Webster. Regular injections of this hormone can prolong the 'peak' of lactation (see figure 3), and boost milk yields by around 25 per cent. To maintain this level of production, the cows need to eat more concentrates, and must inevitably be under yet more physiological stress. This is likely to mean increased susceptibility to infection, and even earlier culling of 'spent' cattle. In a situation where milk quotas are already required to keep dairy production down, the benefits to farmers, consumers or anyone apart from Monsanto, who produce BST, are difficult to imagine.

In the United States a new style of hormone growth promoter for pigs has also been attacked on welfare grounds. The drug cholecystokin is based on the pig's appetite hormones, and when injected at regular intervals it blocks out the natural controls which tell the pig when it is full up. 'Basically these pigs are going to be transformed into eating automatons', claim the Humane Society of America. 'Their whole nature will be fixated by drug-treatment on just one very narrow form of behaviour: eating'.[41]

Another area of welfare concern in both traditional and modern farming is the practice of routine mutilation of animals by operations such as castration. Other common surgical procedures include the debeaking of poultry, dehorning of cattle and sheep, tail docking and tooth clipping of pigs, and various forms of punching and branding for marking purposes. Traditional husbandry often relied on castration, rather than physical separation, as a means of avoiding indiscriminate breeding. The practice is controlled by various regulations, and if properly carried out at an early enough age should not cause too much pain. With intensive systems, and where the male animals are killed before maturity so that there is no question of the meat becoming unpleasantly tainted, there is less reason for castrating. In fact, it can be disadvantageous as growth is often set back; entire animals also tend to be leaner. Surgical castration is forbidden for poultry in the United Kingdom, though chemical methods have been employed to produce capons by altering the hormone balance. Chemical or immunological techniques are a possible future welfare advance for other species, though consumer resistance might be expected to meat thus treated.[42,43]

Debeaking of poultry is carried out to reduce the danger from birds pecking each other. It involves burning off the last third of the beak, and may be carried out as a routine practice with chicks of about two weeks old, or as a response to cannibalism in adult birds. Overcrowding and competition for resources are probably the main causes of feather-pecking and cannibalism among poultry, but dietary deficiency may be a contributory factor. A recent British survey showed that 28 per cent of caged layers were regularly debeaked, 74 per cent of breeder hens and 92 per cent of turkeys.[44] The Farm Animal Welfare Council recommend that beak trimming of hens should be carried out 'only as a last resort', but accept that it is essential for turkeys kept in daylight conditions, which 'can be vicious'.[45] It is sometimes claimed that debeaking is similar to clipping a human's finger nails, and causes the bird no harm. In fact, the beak contains a complex arrangement of nerves, soft tissue, horn and bone to within a millimetre or so of the tip. Recent research indicates that the pain experienced by debeaked hens is long lasting, and can be compared to the post-amputation ('phantom limb') pain felt by many human amputees.[46] Regrettably, some free range flocks are now being debeaked, though the mutilation is forbidden in flocks which conform to the Soil Association standards.[47]

Like beak trimming, the practices of tail docking and tooth clipping in pigs are responses to husbandry problems which would be far better avoided than countered by mutilations. They are usually carried out within the first few days of life, and tail docking is subject to regulations aimed at reducing the amount of pain inflicted.[48] The removal of antlers in velvet from deer is generally prohibited in the United Kingdom, but they may be removed as soon as they are out of velvet, for the safety of farm workers and other animals.[49] The same reasons can be used to justify dehorning of cattle and the trimming of boars' tusks. Not all novel forms of mutilation are responses to the problems of intensification; the operation of 'mulesing' which is carried out on many millions of Australian sheep every year is intended to reduce the likelihood of fly strike in the extraordinarily wrinkly skinned Merino sheep. When blowflies lay their eggs in the moist folds of skin near the sheep's tail, maggots can burrow into the flesh and cause suffering or death. The primitive surgical technique which is used to reduce the chance of fly strike involves cutting a chunk of skin from each side of the tail, so that when the scar heals the skin is drawn taut, and the likelihood of moisture collecting to give suitable conditions for maggots is thus reduced. The operation was devised in the 1920s by J. H. W. Mules, as 'improved' strains of Merino with ever-thicker wool and more wrinkled skin became increasingly afflicted by strike; and sixty years on a more humane solution to this problem is still awaited.[50]

Regardless of the husbandry methods adopted, all farm animals are ultimately destined for slaughter. Many small abattoirs, often with somewhat primitive facilities, closed down during the period of meat rationing in the 1940s, so that most of the 33 million cattle, sheep and pigs and 470 million poultry slaughtered each year in the United Kingdom now go to just about 100 large slaughterhouses. In many respects the handling of stock at the time of slaughter is more efficient than it used to be, but in a highly competitive industry where profits are slim, bankruptcies not uncommon, and piecework pay the norm, the humane objectives of legislation are not easily realized.[51–53]

There are two stages to slaughtering. First, the animal is rendered unconscious by electrical stunning or by using a captive bolt pistol which fires a metal rod through the skull and into the brain. An electrified water bath is used for stunning poultry; electric tongs are applied to the head for pigs and most sheep, while the captive bolt pistol is mainly used for cattle. After stunning, the animal has its throat cut; and as blood drains out of the body the heart stops, and death occurs. If stunning is not carried out effectively before the throat is cut, it may take several minutes for unconsciousness to occur through loss of blood.

In practice there are a number of factors which tend to make electrical stunning unreliable, and there is an urgent need for improvements in equipment and operating standards.[54,55] Even when stunning is effective,

Modern slaughterhouse in Britain. (Reproduced by kind permission of Katharina Reifschneider)

consciousness is often regained after a brief period, so that short delays on the slaughter line may result in increased suffering.

The Slaughterhouse Act of 1974 prescribes that all animals which are killed for food must be stunned before they have their throats cut and are bled out, except in the case of religious slaughter by Jews and Moslems, who are exempt from this requirement. There are slight differences between the Jewish *Schechita* method of slaughter and the *Halal* practice of the Moslems; but basically both involve cutting the animal's throat with a sharp knife without any pre-stunning. For cattle, a restraining device known as the Weinberg pen has been used. This consists of a strong crate into which the animal is fastened before being slowly turned upside down to have its throat cut. The stress and fear experienced by the beast must be considerable, and it has been estimated that several minutes may elapse between the blow of the knife and loss of consciousness.[56] In 1987 the Ministry of Agriculture at last took steps to ban the use of the Weinberg casting pen, and to insist on the use of an approved type of upright pen instead. After slaughter, up to a third of Schechita-killed animals fail to be declared Kosher, and since there is also a religious prohibition on the consumption of certain veins, lymphatic vessels and nerves, the entire hindquarters are also invariably rejected. Consequently, about three-quarters of Schechita-killed meat is sold on the non-Jewish market, without any indication that it has been religiously slaughtered. Nor can it be guaranteed that Halal meat is destined exclusively for the Moslem market.

Under ideal conditions, ritual slaughter may be at least as humane as practical in many slaughterhouses where pre-stunning is not effectively carried out, but restraining the animals is inevitably more difficult. The consensus of scientific opinion outside the ethnic communities is in favour of pre-slaughter stunning.[57] Some Halal slaughterhouses in Britain now stun the animals before slaughter, as a number of Moslem religious leaders have accepted a re-interpretation of a ritual which was originally intended to provide a relatively painless death.[58] Many Jews, too, have ceased to insist on Kosher meat out of concern for animal welfare. Unfortunately, other members of these faiths have taken the recommendations of the Farm Animal Welfare Council, the RSPCA, and other animal welfare groups, that stunning should be obligatory, as an attack on their religious freedom. But do even the hardest line traditionalists among Jews insist that their meat comes from animals that have been allowed free range on the Sabbath, as recommended by the Talmud; or that the eggs they eat have not been removed while the hen was in sight? Perhaps there is disagreement over how far freedom of belief implies freedom of action, even to the detriment of other members of the community; or perhaps they simply cannot acknowledge that animals are in any sense members of our community, or have interests that are worthy of consideration.

Although abattoir slaughter is prescribed for most animals intended for human consumption, deer may be shot in the field, which is reckoned more humane than subjecting the beasts to the stresses of transport and handling. The situation is somewhat uncertain at present, and if deer farming continues to expand it is probable that there will be increased demand for abattoir slaughter on hygienic grounds. Veterinary inspection of game meats, as required by a forthcoming EC directive, could remove some of these objections to field slaughter, but would also make it less economic. And while detection of diseased animals would probably be easier under slaughterhouse conditions, the potential for cross-infection on a large scale is enhanced.

Killing of fish is not regulated at present, though the Farm Animal Welfare Council are currently considering the slaughter methods used at fish farms. These include knocking out each fish by a blow to the head, stunning using carbon dioxide gas bubbled into tanks, or bleeding the fish alive which is said to improve the flesh quality and to be favoured by suppliers to the French market. Slaughter of non-food animals such as mink can be carried out humanely by a barbiturate injection, though many cruder methods are doubtless used as well.

The suffering of animals at the time of slaughter is preceded by the traumas of transport and market day. As road transport has become relatively cheaper and slaughtering more centralized, the distances over which animals are moved have become greater. When live animals are handled by the lorry load, or worse still by the ship load, it becomes a practical impossibility to attend to their needs individually. Most unceremoniously dealt with of all are poultry on their way to slaughter, particularly spent laying birds. They are usually packed into plastic crates, and stacked on flat lorries in loads of around 3000 birds per vehicle. The design of crate in common use has been criticized as difficult to rope securely, particularly when the plastic becomes brittle with age.[59] Environmental control is impossible on the lorries, with the result that in cold weather the birds on the outside may freeze, while in warm conditions some of those inside are likely to suffocate. The length of journey varies considerably, because prices may be better at a pet food factory 250 miles away than at the local soup or baby food plant. In one case which came to court recently, fifty-eight birds were dead in their crates, and over 60 per cent of the load had one or more broken wings or broken legs.[60] In the United States, the majority of broiler operations are 'vertically integrated', with rearing sheds, feed mill and processing plant all on one site. This arrangement avoids road transport of live birds, though it encourages the use of offal, poultry manure and diseased corpses in the formulation of feeding stuff. In Britain such vertically integrated plants are less common, though Unigate have recently built a 350,000 bird broiler complex on Tyneside. An estimated 185,000 lorryloads of hens are still moved around the country each year on their way to slaughterhouses.[61] Under European Community regulations, a maximum journey time of twelve hours without access to water is permitted; and following complaints by the RSPCA that Britain was failing to comply with EC legislation, the government introduced an order which at least obliges lorry drivers to keep records of journey times.[62,63]

Most sheep and cattle traditionally pass through livestock markets, and so are involved in at least two road journeys and in some cases many more. Complaints that calves were sometimes going through five or six markets in as many days prompted legislation to forbid cattle under five weeks old from being sent to market more than twice.[64] The welfare of animals at markets is the subject of a report by the Farm

Animal Welfare Council, which makes a number of recommendations for improvements, particularly in the smaller and more old-fashioned markets.[65]

The most appalling problems regarding animals in transit relate to the growing international trade in livestock. Long journeys in which extremes of temperature and unexpected delays are often encountered make the trading of livestock from country to country the cause of much suffering. It is often difficult to implement regulations on feeding and watering even where these are agreed between states, as in the European Community. Sheep are still transported on three-deck open-topped lorries, where those on the top deck are exposed to rain and cold, and the risk of damage from overhanging trees. But the suffering of sheep on their way to Europe is nothing to what those exported live from Australia and New Zealand have to endure. To its credit, New Zealand had banned live exports after over 4000 out of a cargo of 30,000 sheep sent to Iran in 1973 failed to survive the voyage; but in 1985 the decision was reversed after prolonged lobbying by the farmers and the Australian-based Elders IXL group.[66] The sheep destined for export have often been trucked long distances without food or water, and are then kept for several days in feedlots before being loaded.[67] On the journey to the Middle East, they are packed three to the square metre for an eighteen-day voyage, and after unloading they are kept in holding yards before going to the slaughterhouse. Substantial mortality occurs at every stage of the journey: in 1983 15,000 sheep died of exposure in an Australian feedlot, and in 1981 over 12,000 died on board the *Persia* due to mechanical breakdowns. And when the *Farid Fares* caught fire and sank off South Australia in 1980, more than 40,000 sheep were either drowned or burnt alive.[68]

Their sheer scale makes accidents of this sort all the more horrifying. With the modern trend to larger and larger units, mishaps such as fire, flood or mechanical breakdown are often the cause of disaster on a large scale. During the August 1987 Bank Holiday, 70,000 layers out of a flock of 90,000 birds owned by the New Dawn group suffocated in their cages because of a failure of the ventilating system in the brand new poultry house.[69] Nor is this an isolated incident, but simply a worse than normal example of a pattern of accident which is being repeated all the time. Fire has always been a hazard on farms, but the scale of modern livestock houses, and their frequent remoteness, magnify the risk. Even if the alarm is raised promptly, the problems of evacuating a large animal house quickly enough to save the stock may be insuperable. In a single poultry farm fire in 1986, 24,000 birds were either burnt alive or asphyxiated; and a total of over 60,000 farm animals died in fires during the course of the same year.[70] As with domestic fires, the causes often involved faulty heaters or electrical equipment. However large the livestock unit, there are at present no mandatory fire precautions to protect farm animals (in contrast to the requirements of the Zoo Licensing Act 1981 or the Animal Boarding Establishments Act 1963); nor are there any financial inducements to encourage farmers to take fire precautions more seriously.

In all farms, from the most modern to the most traditional, animal welfare is still dependent to a large extent on conscientious stockmanship. A good stockman will tend to anticipate and avoid problems, while lazy, incompetent or vicious husbandry will inevitably aggravate them. Farmers who remember the bad old days are constantly pointing out that while old-fashioned methods can work well in skilled hands, they could also be disastrous. Many intensive livestock systems were introduced to avoid the difficulties experienced on traditionally run farms where the farmer either could not get skilled labour or was unwilling to pay for it.

As a way of keeping animals warm, relatively clean and adequately fed with the absolute minimum of human effort, there is a lot to be said for wire cages, conveyor belts, concrete slats, slurry pits, thermostats, fans and the other paraphernalia of modern farming. But the need for care and attention does not disappear altogether, as this RSPCA report shows:

> [The] manure – from 17,500 birds – had simply been left to accumulate (for more than two years, it later transpired) and was between four and five feet deep, and in places even deeper. The large double doors at the ends of the sheds had been burst open by the sheer weight of the slowly moving sea of wet excrement.... Birds which had fallen from the rusty and broken cages above had perished in the mire, and there were live birds trapped in it and still struggling to escape.[71]

Even assuming the factory farm can be kept running efficiently, though, it is increasingly apparent that there is more to animal welfare than a semblance of cleanliness and satisfactory feed conversion ratios. And however much the mechanism of the high tech farm can be adjusted to give a reasonably satisfactory environment for the ideal 'average' farm animal, it has to be faced that the real livestock do not always perfectly conform to their roles as such parts of the agricultural machine. To those who know them, the seemingly identical sheep or chickens in a flock have their own individual personalities, and the essence of good stockmanship is to know the animals as individuals. Such knowledge enhances the ability to diagnose sickness, and can also improve productivity. In *Tess of the d'Urbervilles* Hardy discourses at length on the idiosyncrasies of dairy cattle, in a manner which would seem fanciful to many people brought up to assume that farm animals are more or less interchangeable replicas of each other. But there is scientific evidence that cows themselves recognise fifty or more other cows as individuals, and even the most up-to-date dairy farmer knows that he cannot afford to neglect the differences between his animals – or between the various humans who look after them.[72] Stockmanship is also an individual trait, which is difficult to analyse, but important none the less. One farmer claims he can tell which stockman looked after a veal calf by its performance, or even from the quality of its carcass; another relates that by moving his employees from farm to farm he discovered that a good stockman can improve milk yields by up to 20 per cent though using ostensibly the same routine.[73]

The real tragedy of modern agriculture is not simply its scale – though the prospect of hens choking to death by the ten thousand seems somehow more horrific than any number of such incidents in flocks of ten or a hundred – but its impersonality. Escalating flock sizes have not been matched by increased workforces, but quite the reverse, and as a result the number of 'contact hours' with its keeper enjoyed by each animal has been drastically reduced. However efficient the farming machine has become, it is impossible in the larger units for this efficiency to wear a human face. And if the animals cease to be recognized as anything more than cogs in a productive machine, to be discarded and replaced when they wear out, such an attitude not only hurts them: it threatens to destroy our own humanity.

Part IV
THE ENVIRONMENTAL ANGLE

10
Livestock in the Countryside

An old man in the crofting village where I live had a daughter who had moved away to the mainland. On one of her visits home, she brought her parents a goldfish in a bowl. When her father realized the fish was never going to reach edible size, he was vexed. 'Then where', he asked, 'is the profit in it?'

Hard times make sense of crude economic arguments, but when unremitting toil is no longer the order of the day, people become concerned with values beyond mere survival. There can be few readers of this book who could share the old man's genuine incomprehension, but the protestant work ethic which his attitude epitomizes, in which work is the principal virtue and utility the chief end, is none the less an important element of our cultural background. It is sometimes only too easy to be convinced by the sort of economic reasoning which assumes that profit is the ultimate, if not the only goal that is worth pursuing.

But is society nothing more than an economic machine? And is the countryside simply a great factory producing the fuel on which that machine runs? If not, then we must have other concerns beyond the quantity or even the quality of the food that each acre can produce. I have already argued that there are reasons outside the scope of economics why we should care about how farm animals are treated, and it may be that we have a similar sort of duty to care for the countryside itself. Of course, the fields and hedgerows cannot suffer in the same way as sentient beasts – though their wild inhabitants may be injured by the plough, the reaper and the spray – but to even the most urban of us the country landscape has a value not easily translated into money terms.

The sentiments aroused by lush meadows, tinkling streams and dancing butterflies are the perennial stock-in-trade of poets and landscape artists; and it would seem that the countryside has something to offer all our different moods. The pastoral ideal was a nostalgic evocation of innocence, in which actual memories of childhood were freely mixed with the vision of a pre-industrial earthly paradise. Its appeal today is as strong as ever it was to Horace's Roman audience: witness the exploitation of rural images to advertise cars, chocolates and all manner of other consumer goods. The Romantic poets sought and found God in the sterner but more exhilarating face of wild nature. Such attitudes hardly reflect the feelings of the down-to-earth countryman, who might dismiss them as urban in origin and essentially impractical. This is not to say that country-dwellers are devoid of feeling for their surroundings, but their attachment tends to be more particular and local rather than general and aesthetic. Just as aboriginal people identify with their homelands and find it impossible to move without losing an essential part of themselves, the British rural population is tied to the land by centuries of history and tradition. Their affections tend to find expression in corny patriotic verses, or popular songs about the White Cliffs of Dover; which might not count as art for the sophisticated reader of the *Georgics* or *Khubla Khan*, but this does not devalue the underlying sentiment.

However much we love the country as it is, can we necessarily argue that all changes are for the worse? The entire landscape of lowland Britain has been transformed by agriculture, and the natural forests and swamps largely replaced by arable land and pasture. In the seventeenth and eighteenth centuries landowners could defend their reconstruction of the countryside as 'improvement', and often devoted great energy to laying out their estates in conformity with the aesthetic fashions of the period. Today, we have none of their confidence, and little wonder. Certainly it is unusual to hear the erection of a new battery house, the ploughing up of permanent pasture, or the dynamiting of ancient oak trees defended on grounds other than the pursuit of profit; nor is it easy to think of recent agricultural innovations that could be said to have had any benevolent impact on the rural environment. In the post-war period the impact of mechanized and specialized agriculture on the landscape has been staggering: hedgerows have been removed at a rate of up to 10,000 miles per year, and some counties in the East of England have lost 80 per cent of their trees.[1] If the rest of this chapter concentrates on the negative environmental effects of intensive farming methods, it is because even the farmers themselves seem unable to find much to say in their favour. The idea of 'improvement' is as unfashionable with developers as with their opponents; and today's watchword is 'conservation'. Current attitudes tend to presume that imposed changes are invariably harmful to the environment, and if the status quo which has evolved over the past centuries is to be disturbed, the best we can hope for is to limit the resulting damage.

While the number of people in rural employment has declined steadily over most of the past two or three centuries, there has been a corresponding increase in appreciation of the recreational value of the country, an explosion in the number of city workers travelling daily or at weekends to country retreats, and more recently a slight increase in rural population even in the more remote areas. The aesthetic impact of modern intensive farming has therefore been of concern to a growing number of articulate users of the countryside, who tend to object both to the obtrusiveness of modern methods and to the disappearance of their more picturesque precursors. Visually, factory farms are as unattractive as most other industrial developments, having been erected under the same constraint of minimum expense, and with even less in the way of planning controls. Since there is a far greater area of agricultural land than of land zoned for industrial and other purposes, the impact on the countryside of uncontrolled jerrybuilding by farmers has been very widespread. Perhaps even more noticeable, though, has been the change from mixed farming to monoculture, and the disappearance from the fields of the traditional farm animals, even in livestock producing areas.

The industrial process, whether the end-product is food or television sets, can be described in terms of a closed box into one end of which are fed certain raw ingredients, and from the other end comes the finished goods. What goes on inside the box is something of a mystery, and even when it is displayed it remains too complex to grasp for the average observer, as for the average production-line worker. The old-fashioned farm, now so scarce that it can prove a successful tourist attraction, is something of an antidote to the mystification induced by the industrial society, as it functions on a scale we can comprehend. When the countryside too becomes industrialized, the alienation of the average person from meaningful productive work becomes even more complete. The different slant on country life given by increasing industrialization can be demonstrated by examining the rural ideal depicted by seventeenth-century English poets. In *The Politics of Landscape* James Turner points out they hardly refer to pruning, digging, weeding, thatching, hedging, 'almost everything which anybody *does* in the countryside is taboo'.[2] Contrast this with the current obsession for rural and small-scale crafts, and the widely felt desire to get back to the land as a meaningful work experience. Your modern drop-out does not just want to play at shepherdesses like Marie Antoinette, but actually wants to *be* a shepherdess in the quest for an existentially authentic life style.

Quite apart from the disorientation which radical changes in agriculture may or may not induce, to many people there is great pleasure in watching a field of sleek and contented cattle and seeing the breeze ruffle the clover and vetches of an ancient meadow, which cannot be replaced by contemplation of a corrugated iron byre, or the vista of endless acres of silage grass. But because no-one has yet devised a way of measuring such feelings they are largely ignored in efforts to regulate the nuisance caused by intensive livestock units, which concentrate instead on the more quantifiable offence given to the senses of hearing and smell.

One of the worst nightmares I ever had occurred in the pretty thatched inn of a charming and remote Devonshire village. I had gone to sleep with the muffled sounds of rustic merry-making in my ears, but my dream transmuted the happy talk into hideous and orgiastic cries, accompanied by suitably Breughelesque visual imagery. The horror was that the more I tried to wake, the more real the Bacchanalian shrieks became, until eventually and shaking with fear, I traced the screams of the thousand damned souls to the piggery a couple of fields away. What went on there, I still don't know, but quite possibly if one lived there for a little while one would cease to notice the racket at feeding time, the noise of the ventilation fans and the coming and going of heavy vehicles bringing fodder and taking away baconers. For many unfortunate occupants of once peaceful hamlets such constant disturbance is now the grim reality, and if it does not give them bad dreams, sometimes it must surely stop them sleeping at all.

More pervasive, and a more frequent cause of complaint than noise, are the offensive smells which emanate from intensive livestock units and their associated agricultural practices. There are currently over 2000 complaints per year in England and Wales about farm smells, and almost all relate to intensive animal husbandry. People living near large livestock units are exposed continuously to the ventilation exhaust gases and smells from slurry tanks, and intermittently to the release of noxious vapours as buildings and tanks are cleaned out. And since slurry is generally disposed of by spraying it on to the land, people over a much wider area are then exposed to powerful and unpleasant smells. While many farmers would like to explain away the increasing number of complaints about odour pollution as emanating from a new generation of 'yuppie' country dwellers, Peter Roberts points out that it is not the country dwellers that have changed so much as the smells.[3] When animals were bedded on straw, manure was accumulated in dung heaps where fermentation in the presence of air maintained a high temperature which destroyed pathogenic organisms. In spring, the manure, which had a 'good country smell', was returned to the land as a fertilizer and soil conditioner. In contrast, the slurry of faeces and urine which is produced in animal houses where no bedding is provided requires great care in handling if it is to be disposed of safely. The normal procedure is to store the slurry in tanks or lagoons where it decomposes in the absence of air, producing a number of very strong smelling substances such as ammonia, hydrogen sulphide, mercaptans and skatoles. Harmful micro-organisms are destroyed during this anaerobic fermentation, but less efficiently than in the heat of the traditional dung heap. Whenever the slurry is disturbed, as during spreading, the strong smelling products are released, though once it is spread on the land it gradually oxidizes and the smell disappears.

Apart from the sheer unpleasantness of the smells associated with intensive livestock houses and slurry disposal, there may be actual health hazards to people living in the area. The risk to human health is uncertain, but there is evidence from studies of viral epidemics affecting farm animals that contagious particles can spread on the wind up to 30 kilometres from the site of infection.[4] Dust from the ventilators of intensive livestock units may be responsible for the spread of disease to animals some distance away; and farmers are advised not to apply slurry to the land by spray gun, since droplets can drift for considerable distances and might infect grazing animals.[5]

A further nuisance indirectly associated with intensive farming is the smoke and fire risk caused by the annual burning of around 5 million tons of straw. The increase in barley production, and a reduction in demand for straw as bedding because slatted or solid floored animal houses are perceived as more efficient, have led to a huge surplus of straw. Ploughing it in is costly, and as well as increasing the nitrogen demand on the soil, the decomposition process can produce toxins inimical to the growth of the succeeding crop. In modern specialized farming, the straw from cereal growing and the droppings of housed livestock have come to be regarded simply as waste products to be disposed of as cheaply as possible. This contrasts with the traditional mixed farming approach, where the manure produced from straw and dung was the mainstay of soil fertility: less than 50 years ago tenant farmers could be evicted if they sold straw off the farm, let alone burnt it.[6] After a long campaign fought by countryside environmental groups, the government has at last announced an end to straw-burning from 1992.

Modern farming methods have a more widespread environmental impact than simply being a nuisance to oversensitive neighbours. In the United States it is claimed that there is more pollution from agriculture than from all other industrial sources, mainly by contamination of water courses.[7] In many parts of Britain there is serious concern over excessive nitrate levels in drinking water, due to a vast increase in the use of nitrogenous fertilizers. Between 1961 and 1980, nitrogen consumption per hectare of agriculturally used area increased from 24 kg to 72 kg in the United Kingdom, and from 24 kg to 78 kg throughout the European Community.[8] Although much of this increase is related to more intensive cereal cultivation, it also reflects the move from permanent pasture and hay crops to zero-grazing and silage production as a more efficient use of grassland. Thus while the acreage of grass has decreased, the number of livestock has grown – but so have capital and labour costs, and the reliance on bought-in concentrates to supplement the animals' diet. Even with highly soluble fertilizers such as ammonium sulphate, leaching losses are far lower from land which is covered with a growing crop, so pollution by nitrates is mainly a problem of arable regions.

In the western counties of Britain, intensification of beef and dairy farming has led to a major problem of water pollution by farm wastes. In the past ten years farm pollution incidents have more than doubled, with more than 75 per cent being associated with cattle farming.[9] About a quarter of all farm pollution incidents involve silage effluent. The anaerobic fermentation which occurs during silage production produces – particularly if the grass is wet – a waste liquid up to 200 times more polluting than domestic sewage. (Very little effluent is produced if the moisture content of the crop is less than 75 per cent, but at 84 per cent moisture, each ton of feed crop gives rise to over 100 gallons of effluent.[10] If this waste is allowed to enter rivers or streams the natural aquatic life may be poisoned, or suffocated by the reaction of the waste liquor with essential oxygen. Silage production trebled in England and Wales during the ten years to 1987, but there is some hope that improved farm practice may reduce the danger of pollution. It is possible to produce silage with a lower water content by allowing the crops to wilt before ensiling, though this depends on suitable weather; and changing from the use of large silos to fermentation in plastic wrapped bales would reduce the risk of large scale spillage.[11]

Given the trend towards larger and more intensive livestock farms, water pollution by animal excrement is a more intractable problem. The polluting potential of farm animals in the United Kingdom, measured in terms of the biological oxygen demand of their excreta, is equivalent to the sewage produced by 150 million people, or nearly three times the current British population. In the United States of America the ratio of farm animal to human 'waste' is more like ten to one.[12] It cannot be overemphasized that it is misleading and unwise to regard animal excreta as intrinsically waste products, simply because in the context of modern livestock farming their disposal sometimes involves expense which the farmer would sooner be without. When farm animals roamed the fields, or were kept inside on straw, their droppings and urine provided valuable fertilizer with minimal risk. But since an estimated 60 million tons of undiluted excreta are now voided indoors, compared with 110 million tons deposited over the entire area of Britain by grazing animals; and since most farmers have abandoned straw in favour of labour-saving hard floors and slurry handling systems, we are now faced with massive amounts of extremely noxious material which presents such a disposal problem that it is natural to regard it as waste. The bulk of slurry is, in fact, returned to the land, making use of its nutrient value for growing crops; but there are many problems associated with its storage and application. About half of all farm pollution incidents involve spillage from slurry stores or contaminated water that has been used to hose down animal houses or for yard washing. There is an additional pollution risk when the slurry is actually spread, since if the land is waterlogged, or if more nutrients are present than the growing crops can use, the surplus material can run off the land into drainage channels and other waterways.

The safe disposal of waste material is a challenge currently facing many industries, as it becomes increasingly evident that make-shift methods used in the past have unacceptable long-term consequences. In the traditional mixed farm nothing was wasted, and there was little need to import fertilizers or feeding materials. The availability of cheap artificial fertilizers, increasing specialization among farmers, and the substitution of expensive labour by oil and electric powered machinery, have changed the nature of the farm from a relatively self-sufficient unit of production to a consumer of raw materials and a generator of waste.

In dairy and beef farms, where the livestock are still basically fed on grass, even if this is supplemented by bought-in feeding stuffs, there is usually sufficient area of grassland to absorb all the animal excreta as useful fertilizer. While handling faeces and urine as slurry is less labour intensive than using straw bedding, it creates a whole new set of difficulties. Because slurry spreading cannot be carried out if the ground is too wet, six months' storage capacity must be provided, which can amount to half a million gallons or more.[13] The environmental impact of such a store failing could be disastrous. The anaerobic fermentation of the slurry in its storage pond produces large amounts of noxious gases which create a smell problem, and also represent a loss of potentially valuable nutrients to the atmosphere. It has been estimated that the escape of ammonia, etc., from open slurry stores can reduce the nitrogen content of the material by up to 60 per cent.[14] More important, there are significant risks to animal and human health from the use of slurry. With normal bedding, three weeks in the dung heap will produce a temperature high enough to disinfect the material fairly thoroughly.[15] In liquid manure, salmonella can stay alive for 350 days at winter temperatures, or 180 days in summer; and more than ten years ago tests on cattle slurry showed the presence of salmonella in 20 out of 187 samples.[16,17] To minimize the risk of infection, farmers are officially advised not to spread slurry on land which is to be grazed, or if this is not practicable, to store the slurry for at least a month before spreading, and then to leave the pasture ungrazed for a minimum of six weeks.[18] The possible presence of pathogenic organisms in farm waste water and slurry adds to the dangers attending their escape into river water. Bacteria can multiply by a factor of 100,000 in river water containing a hundredth of one per cent of organic substances, and contaminated waterways can transmit infections such as salmonella, *E. coli* and leptospirosis to other animals or to humans.[19] Chemical disinfection is sometimes proposed as a solution, but while small amounts of waste can be treated with lime and applied without hazard to croplands, use of chlorine or other chemicals creates further pollution risks and cannot be recommended.[20] Vigorous aeration will also disinfect slurry, but involves expensive equipment and the use of energy: attempts to achieve sufficient aeration by natural flow in ditches and lagoons are unsatisfactory, especially in cold climates.[21]

The huge local concentration of animals in modern pig and poultry farms (or in the beef feedlots common in Eastern Europe and the United States) presents a major challenge of waste disposal. Such intensive holdings seldom have sufficient land of their own to absorb the waste they produce, and in regions where there are large numbers of intensive livestock operations the cost of transporting slurry to areas where it can be spread may be substantial: one author quotes 3 per cent of production costs for trucking waste a distance of 5 km.[22] One hectare of wheat can only make use of the nutrient elements present in the waste produced over a year by 17 pigs or 300 laying hens; though it may be possible to apply up to double this amount of excreta without actually reducing crop yields or causing gross water pollution.[23,24] Since some factory farms accommodate up to 100,000 pigs or several million laying hens, waste disposal can easily become the limiting factor which determines the maximum viable size of the unit. In Hungary, problems of waste management have necessitated a reduction in the recommended size of pig farm to no more than 10,000 animals; and while Russian beef feedlots have contained up to 30,000 beasts, the Czechs have restricted their size to not more than 1000 cattle in the interests of safe excrement control.[25] In Sweden, too, the density of animal population is restricted according to mean nitrogen disposal requirements. Nearer home, the concentration of intensive livestock farms in The Netherlands has put great pressure on land in that country. In January 1987 the Dutch authorities imposed a ban on further expansion of livestock numbers. With farm effluent production running at 100 million tonnes per year, they already have twice the amount that can safely be spread on the country's 2 million hectares of farmland.[26] A system of manure banks has been set up, under which the transport of liquid manure is subsidized to encourage its use on land away from the factory farms. The practice of dumping slurry on 'sacrifice areas' of land at concentrations of up to 100 times the maximum beneficial level is likely to cause serious pollution of groundwater, and has now been banned in Holland.[27,28] Dutch farmers must now pay the full cost of disposal through sewage plants if they cannot find suitable land on which to spread their slurry. In the United Kingdom, the concentration of pig farms around Humberside places a severe strain on the surrounding arable land, where a number of sacrifice areas have been created, and water pollution incidents continue to increase.[29,30]

Further environmental risks which have been attributed to the use of factory farm wastes as fertilizers include contamination of land by additives used in animal feedstuffs (e.g. copper salts given to fattening pigs), or by antibiotics.[31] The soil surrounding large animal houses is heavily contaminated with antibiotic resistant micro-organisms, and fears have been expressed that this resistance could be transferred to other families of soil bacteria.[32]

In the hope of avoiding the unpleasant and potentially dangerous effects of slurry spreading, a number of alternative schemes for treatment of animal wastes have been devised. All are aimed at making use of the valuable elements of the waste materials, and simultaneously reducing their offensive nature. The end product may be a plant nutrient, animal feed or energy generated by using the manure as fuel.

The unpleasantness of slurry can be reduced, and its handling made easier, by mechanically separating the solid and liquid fractions. The solid element is stackable and not too smelly, while the clarified liquid can be pumped to the land by irrigation pipes. Mechanical aeration is an alternative way of deodorizing slurry prior to its use on the land. High quality compost can be produced by a modification of the old-fashioned dung heap, in which manure with a moisture content of around 50 per cent is aerated in an insulated 'bioreactor' to maintain optimum conditions for decomposition. Composting takes a mere three to five days, but the process only makes sense economically if the end-product can be marketed to gardeners or vineyard owners.[33] There must surely be even more doubts as to the viability of a process described by the Royal Commission on Environmental Pollution, in which pig slurry is mixed with straw before composting in specially constructed chambers.[34] If straw is available, it would surely make more sense to give it to the pigs as bedding, and let them do the mixing.

A more controversial use of animal faeces is the production of feeding stuffs for livestock of the same or different species. Unless the most rigorous sterilization procedures are adopted, this practice carries a high risk of perpetuating infections such as salmonella. However thoroughly the animal houses are cleaned out between successive flocks, contaminated feeding stuff provides a simple and common route for reinfection. In theory, ensiling poultry litter for seven days will eliminate salmonella; but heating to 150° C for three hours is the only method of guaranteeing complete sterilization.[35] It is claimed that cattle fed up to 40 per cent ensiled feedlot waste showed 'no detectable adverse effects'; and that milk production is maintained on a diet in which 20 per cent of protein is provided by dehydrated hen manure, but there are said to be 'problems of acceptance' of dehydrated manure by ruminant animals.[36] Since faeces contain about half of any drugs which were in the animals' feed, difficulties may occur when they are fed to other species. For example, cattle suffered abortions when fed on dried poultry manure with over 0.1 part per million equivalent of oestrogens. Similarly, copper toxicity has been reported in ewes fed diets with 25 to 50 per cent poultry litter.[37]

A recurring idea for the use of animal wastes – originally suggested by Pasteur in the 1880s – is as a source of energy. A number of different methods have been tried: pyrolysis to produce oils, tars, gases and charcoal; hydrogenation, which increases oil production at the cost of giving highly contaminated water as a by-product, and anaerobic digestion to produce gases and a residue which can be used as a nitrogenous fertilizer.[38] Such processes can only be efficient if there is a demand on site for the energy they produce, and in view of the high capital and operating costs and the relative cheapness of alternative forms of energy they seem unlikely to be viable at present. Out of over 1000 digesters built in France and Germany during and immediately after the Second World War, only one remained in operation in 1976.[39]

While the potential for pollution from modern agricultural techniques is enormous, strict controls could force farmers to adopt established methods by which the risks can be reduced. If none of the techniques for using the slurry from intensive livestock units is economic, and farmers have to carry the full cost of agricultural pollution control, there may be an incentive for a return to more traditional bedding on straw.

Although the farmer's role as a polluter has been emphasized, it should be remembered that agriculture already plays a valuable part in absorbing pollutants produced by the rest of society. Out of more than a million tons of domestic sewage sludge produced annually in the United Kingdom, around 40 per cent is spread on agricultural land. The rest is used for land fill or dumped at sea, wasting material which could be used as fertilizer. Thorough treatment is required to avoid health risks, particularly in the case of crops that can be eaten raw, and some domestic sewage contains excessive levels of heavy metals or other persistent toxins; but if these problems can be overcome, the technology which has been developed for dealing with agricultural wastes could play an increasingly valuable part in disposing of human sewage waste.

11
Fish Farming and the Environment

A great deal of the environmental impact of modern livestock farming is related to the change from dry fodder and straw-based solid manure to wet feeding and the handling of animal excreta in a moist or semi-liquid state. Not only do the wet substances give off more powerful smells when disturbed, but enormous care is required in their handling to avoid the escape of highly toxic liquids which can easily flow into streams and rivers, or percolate into subterranean water systems. Once such poisons have entered watercourses, there is very little chance of recovering them, or of localizing the damage they cause.

The farmer on dry land has traditionally been expected to accept responsibility for his own waste products, and either confine them to his own land or pass them on by agreement to another party. In the past twenty years, a new type of farming has been established, in which waste food, faeces and chemical residues are silently and effortlessly removed from the farm site to become someone else's problem. Within the constraint of his discharge consent required under the Control of Pollution Act 1974, the freshwater fish farmer simply lets his waste problem drift away downstream, while his marine counterpart operates in a happy-go-lucky state of affairs where he can with impunity dump practically anything he likes into the ocean. In the latter case, there is considerable doubt over what the legal position is under the 1974 Act, and virtually no possibility of effective policing anyway.

Fish farmers rightly point out that their livestock are very sensitive to pollution and require pure water if they are to thrive, with the implication that self-interest will ensure that their activities do not cause any significant pollution. However, their general opposition to proposals for new fish farms upstream of existing freshwater installations, or near existing sea cages, must be based on a belief that there is a certain element of risk attached to these operations.

In freshwater installations (trout farms and the hatcheries producing the one or two ounce size salmon smolts for growing on in the sea) the sheer volume of water required can disturb the river's ecosystem. A hatchery with a quarter of a million smolts requires around 350 litres (75 gallons) per second of pure, well oxygenated water.[1] Until 1988, fish farms were exempt from the system of licensing which controls abstraction of river water for industrial uses, and cases have arisen where the entire river flow has been diverted through the fish farm, leaving the natural river bed completely dry.[2] Even in less extreme situations, the residual flow of the river may be inadequate to protect aquatic life – including important fisheries – or to provide dilution for the fish farm effluent.

The waste water from fish farming contains suspended solids consisting of uneaten food particles, faeces, fish scales and other detritus, and chemicals in solution which have been excreted via the gills or as urine. The most significant effects on water quality are increases in the levels of nitrogen and phosphorus, reduction in availability of oxygen as a result of respiration and in consequence of decomposition of the various solid wastes, and increased cloudiness from suspended matter. Nitrates and phosphates encourage plant growth either as blankets of water weed or in the form of large concentrations of microscopic algae which contaminate the whole water body (eutrophication). Such algal 'blooms' can seriously depress the dissolved oxygen concentration, to the severe detriment of other forms of aquatic life. The effect of increased turbidity is to cut down the biological productivity of the river by stopping light penetration, and as the solid particles settle out they may physically smother river bed habitats.

A recent survey of freshwater fish farms in Britain showed that in only 5 per cent of cases was there clear evidence of harm to adjacent fisheries.[3] In recent years, improvements in feed formulation have reduced the amount of solid waste being discharged, but about half a ton of solids are still released for every ton of fish produced, and with trout production now around 18,000 tons per year, the total pollution load is quite significant.[4] Since nearly 40 per cent of trout farms are sited upstream of drinking water supply intakes, effluent quality has implications for the water supply industry. Concern has been expressed that bacteria harmful to human health can multiply in fish pond sediments or in the fishes' digestive tracts, but the majority of bacteriological studies of fish farm effluents have found little or no evidence for this effect.[5,6] Pollution by chemicals used as medical treatments or disinfectants is a further risk, some such substances being directly toxic and others such as antibiotics possibly allowing the development of resistant bacteria in the river system. The presence of extra solids in suspension increases the cost of water treatment because of the need for additional filtration; and if eutrophication of the water system occurs, the additional treatment needed before it is fit to drink is extremely expensive. Under unfavourable circumstances, pollution from fish farming could cost the water industry up to £6500 in extra treatment costs for every ton of fish produced.[7]

While it has long been realized that bad husbandry can destroy the productivity of croplands, and that industrial or domestic pollution can seriously damage streams and rivers, it is only very recently that we have begun to regard the oceans as in any sense vulnerable to Man's activities. Evidence of overfishing, declining numbers of cetaceans, the damage caused by oil spills and the discovery of toxic chemicals in marine animals as far away as the Antarctic, have all pointed to the conclusion that the seas are finite, both in resources and in their capacity for absorbing the waste products of civilization. James Lovelock observes that the vast majority of biologically productive activity in the world's oceans is concentrated in the shallow waters within a few miles of the coast, which regions he regards as crucial for the maintenance of a stable atmospheric transport system, and hence to the long term fertility of the land.[8]

The rapid growth of the Scottish salmon farming industry, from less than twenty sites with cages in 1980 to over 200 at the present day (see Figure 4), has coincided with a period of greatly increased environmental awareness. The first fish farms were generally welcomed as a sensible use of untapped resources and a valuable contribution to the depressed Highland economy; but within the past few years fish farmers have become the target for criticisms from a wide spectrum of environmental organizations. Such criticism may be on the basis that they interfere with human aesthetic, recreational or economic interests, or because of possible or actual damage to wildlife and the marine ecosystem.

The Scottish Highlands have some of the least spoilt wilderness scenery in Europe, and the large number of tourists who visit the North and West of Scotland are attracted by the unspoilt scenery and peace and quiet. The presence of fish cages in practically every sea loch or sheltered bay is at least as much of a visual blot on the landscape as a major road development or an obtrusively sited electric power line, though the resulting economic benefits to the region are considerably more debatable. As well as the cages themselves, the shore-based facilities for maintenance are often unsympathetically sited and untidy.

Since the most popular activities for visitors to rural areas are visits to coast and lochside, where they can walk, watch birds or simply admire the view, there is the possibility of a conflict of interests between the tourist industry and fish farming. Visitors' enjoyment of the wild

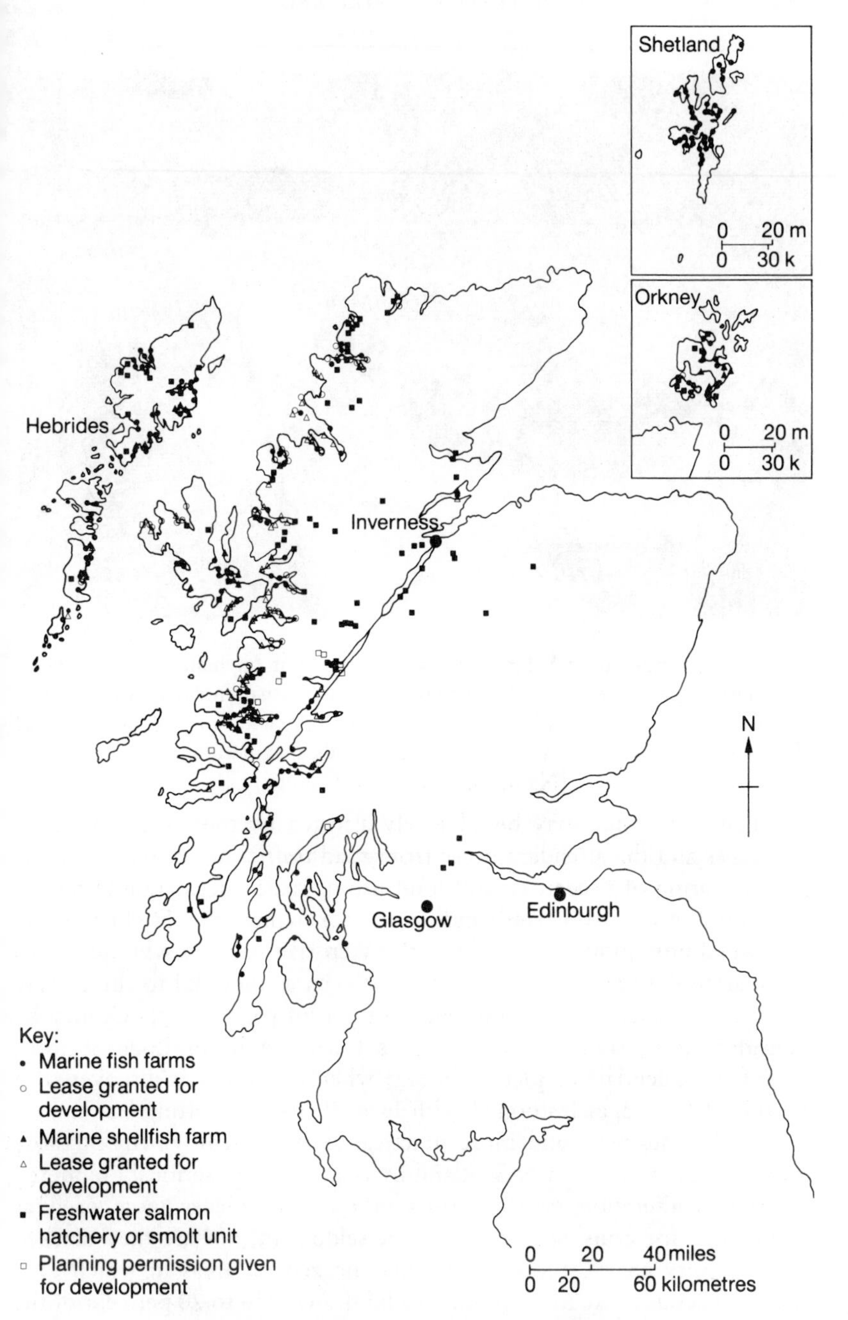

Figure 4 The locations of Scottish fish farms.
(*Source*: NCC (Scotland) *Fishfarming and the Marine Environment*.)

The floating cages of fish farms are a familiar sight on many Scottish lochs. Fish farming has been widely criticized for its visual intrusiveness, for the associated disturbance and harassment of wildlife and for the pollution of the marine environment by synthetic chemicals and natural waste products. (Nature Conservancy Council)

Highland landscape may be adversely affected by the visual impact of fish cages and the attendant noise from road traffic, motor boats, shooting or scaring of predators, and loud use of radios. Remarkably few of the grand vistas over land and water for which the Highlands are renowned now look as they did to the Victorian painters who helped to popularize the region with tourists – or even as they did to the picture postcard photographers of the relatively recent past. The shock may be considerable to someone, who knows Loch Linnhe or Loch Seaforth only from calendars or picture books, when the reality is discovered to be full of fish cages festooned with lurid floats and netting.

At a less aesthetic and more practical level, the enjoyment of those who visit the west coast of Scotland by boat has been seriously hindered by the proliferation of fish farms in the very places most used as anchorages for cruising. Fish cages are seldom lit, nor are they marked on admiralty charts, though belatedly the general areas in which they may be encountered are beginning to be shown. Up to 20 per cent of the best anchorages on the Scottish west coast are now unavailable to yachts or fishing boats because of fish farms – a situation which has implications for safety of life as well as mere convenience. Substantial areas of water which were traditionally used for salmon netting or other commercial fishing activities are now effectively out of bounds for such uses. As a result of this physical interference with free navigation, and because of fears about pollution, the established fishing industry is less than enthusiastic about its new competitor.

For up to a thousand people directly employed in marine fish farming, and more involved in net and cage manufacturing, boat building, feed processing, road haulage and other such related work, the economic benefits of the new industry can hardly be doubted. The jobs created are particularly welcome in that many of them are in remote rural communities where there are few other chances of employment. Concern has often been expressed that with large multinational companies taking up so many of the grants for the new enterprise, the local population are merely being exploited as cheap labour, and are deprived of any real share in the profits accruing. On the other hand, it has been these companies who have funded much of the research, and experience shows that many of the smaller operators (who have been given a degree of preference by agencies such as the Highlands and Islands Development Board) have not been able to make their firms profitable at all, either through mismanagement or simply because of lack of financial stamina. A more serious economic question is whether in the long term, fish farming will not cause such a loss of jobs in industries it impinges upon, like traditional fishing or tourism, that its social benefits are effectively negated by its adverse effects on other sectors of the local economy.

Pollution from fish farming can affect people directly, as when discarded feed sacks or nets are washed up as litter on the shore, or foul the gear of local fishermen. Worse, whole rafts of cages have been abandoned after storm damage or when the operator has become bankrupt, to become unsightly rubbish on otherwise unspoilt shorelines. But mariculture also poses a more insidious threat: the secret and unseen discharge of waste products and alien chemicals into the marine environment can, it is claimed, poison whole ecosystems with devastating effects on wildlife, and indirectly on people who have traditionally exploited or enjoyed the bounty of the sea.

As with freshwater fish farming, an important result of feeding so many fish in a restricted volume of water is to reduce the availability of oxygen for other forms of aquatic life. This is because the fish themselves compete for oxygen, and further demands are made by micro-organisms in decomposing faeces and waste food. Increased levels of nutrients such as dissolved nitrogen can exacerbate the problem by allowing the proliferation of microscopic algae. Certain species of these phytoplankton, such as the notorious red tide forming *Gymnodinium* species, are actually toxic, and can result in death of the farmed fish or pollution of adjacent shellfish stocks to the point where they are dangerous for human consumption.[9]

Under normal circumstances the water within fully stocked salmon cages is depleted of oxygen by up to 2 mg per litre compared with the surrounding sea.[10] In summer this can mean that oxygen levels are approaching the minimum acceptable concentration for marine animals, and it is not uncommon for salmon farms to experience difficulties of 'self-pollution' during warm weather, requiring rapid movement of fish stocks to alternative locations. It might be assumed that if conditions actually within the cages are satisfactory, there can hardly be a problem in the comparatively vast volume of the surrounding sea loch. However, while the surface water where the cages are is generally saturated with oxygen, many lochs have deep basins in parts of the sea-bed, where the water is somewhat stagnant, and the discharge of fish farm wastes into such regions could have a pronounced effect on the deep water oxygen supply.[11] Although the cages are quite small, the biomass they contain is very substantial: if a sea loch were found to contain several hundred thousand fully grown wild salmon, there would be general astonishment as well as much excitement among anglers. To put it another way, pollution by deoxygenation from a single fish farm containing 200 tonnes of salmon is equivalent to the discharge of 12.5 litres per second (10,000 gallons per hour) of untreated domestic sewage – more than most small towns produce.[12]

It is far from being true that fish farm wastes are evenly dispersed throughout the sea before eventually being absorbed. Inevitably, the solid fraction of uneaten food, fish scales, and faeces tends to sink to the bottom directly under the cages. For every 100 tonnes of salmon produced there are approximately 70 tonnes of solid wastes, which can form a blanket covering on the sea-bed of up to a foot in thickness at sites with low current velocities.[13] The high oxygen demand of these sediments results in all normal bottom dwelling life being stifled and replaced by species of bacteria which decompose the organic material, producing toxic gases such as hydrogen sulphide and methane. The presence of such bacteria is indicated by a characteristic furry white growth in the worst polluted areas under the cages. In extreme cases, and particularly when sediments are disturbed during efforts to remove or disperse them, the release of poisonous gases can kill fish in the cages above. Because of this risk, the periodic resiting of cages is recommended. When the cages are removed there is a gradual recovery of the sea-bed, and so far there is little evidence of permanent habitat damage from the organic sediments produced by fish farming. However, since it is impossible to predict the long-term effects, there is a case for prohibiting fish farm developments in the relatively few areas of special marine biological value where it has not already taken place.

Like other intensively farmed animals, caged fish are particularly susceptible to a variety of diseases and parasites, and it is impossible to prevent these being caught from, or transmitted to, the stock in nearby farms or wild fish in the surrounding area. The common practice of dumping dead fish at sea, instead of burying them in lime as recommended, must inevitably assist the spread of infections. Bacterial kidney disease, which is the most important ailment of farmed salmon in British Columbia, has recently shown a dramatic increase in Scottish waters. The bacterium can be caught from infected sites or from wild fish, but the symptoms of the disease do not develop for up to two years, so identification of affected stock is very difficult. When the disease does strike, on transfer of smolts from fresh to sea water, mortality of up to 80 per cent can result.[14] If farms are stocked with fish imported from other locations, new diseases and parasites can be transferred to the native wild fish. In Norway native salmon stocks have been virtually wiped out in a number of river systems by a parasitic fluke *Gyrodactylus salaris*, which spread with the movement of farmed fish from place to place. The only effective treatment is to remove all the salmon from infected rivers and re-stock with clean fish.[15]

A more subtle disturbance of native salmon stocks can occur when large numbers of farmed fish escape into the wild by accident, or are deliberately loosed into river systems in an attempt to boost future catches. Hatcheries produce large numbers of 'S2' parr, which are fish that do not develop into smolts viable in salt water at the end of their first year; and it can be more economical to give these away to fishery managers than to bear the expense of maintaining them for an extra year. The wisdom of this practice has been questioned on the grounds that farm fish tend to perform less well in the wild, but can none the less impair the success of native breeding stocks when introduced in large numbers. Like farmers on land, fish farmers will inevitably try to use selective breeding to give greater docility and rapid growth; and when such domesticated stock is loosed into the wild there is concern that the genetic characteristics of the native fish population will be compromised by interbreeding.[16] In Norway, recognition of these problems has resulted in legislation which prohibits the large scale release of smolts, and the transfer of fish between different river systems.[17]

It has also been suggested that the effluent from salmon hatcheries rearing large numbers of non-native fish can impair the ability of returning wild salmon to locate their home river. The wild fish are believed to rely on the smell of specific pheromones to identify their spawning grounds correctly, and anecdotal evidence supports the view that large scale hatchery operations make this task more difficult.[18]

The extent to which diseases affect farmed fish is a consequence of the crowded conditions under which they are kept, but as with other intensive livestock systems, cure is often cheaper than prevention. To counter the various ailments suffered by fish in cages, fish farmers use a number of antibiotics, insecticides and other chemicals which inevitably disperse into the surrounding marine environment. Antibiotics incorporated in ready-mixed feeds are regularly used for the treatment of such diseases as enteric redmouth and bacterial kidney disease. Already, bacterial resistance to products such as oxytetracycline and trimethoprim is a major problem for the Scottish salmon industry.[19] The drugs prescribed include a number of antibiotics used in human medical treatment, despite the availability of apparently suitable substitutes which are not of medical importance.[20] Overt damage to the environment from these products is unlikely, but as mentioned above there remains the unanswered question of whether or not the widespread use of antibiotics and the development of resistant strains in the natural bacterial population will result in the transfer of antibiotic resistance to pathogens affecting the human population.[21]

The most significant chemical pollution from fish farming at the present date results from the use of the controversial organophosphorus compound marketed until 1989 as Nuvan, but then repackaged as Aquagard. This chemical, which is the active ingredient in fly-kill strips and certain flea sprays for dogs and cats, is used to kill the sea lice which infest salmon cages. Nuvan remains on the Department of the Environment's 'red list' of chemicals they are considering banning, but in June 1989 it was belatedly issued a temporary licence as a medicine in salmon farming, giving the seal of respectability to its widespread but previously illegal use by the industry. (Nuvan was previously available as a pesticide, but could not be used medicinally on live animals without a vet's prescription.) Salmon farmers have repeatedly insisted that without Nuvan they could not continue, given that the alternative organochlorine compounds such as dieldrin, aldrin and chlordane have already been banned. Sea lice can multiply very rapidly, particularly in summer and when cages are crowded or water exchange rates are low – in 1976 all the fish in one of Scotland's first large sea cages were killed by sea lice infestation.[22] Treatment with Nuvan is administered by surrounding the fish cage by a tarpaulin to reduce water exchange, and dosing the enclosed fish at a concentration of 1 part per million for about an hour. The tarpaulin is then removed, allowing the chemical to escape into the surrounding sea. After a week or so, Nuvan breaks down to relatively simple compounds which are not believed to be hazardous to marine life, but unfortunately before this occurs its selective toxicity to arthropods poses a real hazard to any lobsters or crabs in the vicinity. Lobsters are killed by exposure for a few hours to Nuvan at a tenth of the working concentration, and little is known about its possible effects at lower doses.[23] A concerted campaign by conservationists for a ban on its use has been supported by the Clyde Fisherman's Association and the Shellfish Association of Great Britain.[24] The lack of effective regulation of fish farms is illustrated by the fact that the Highland River Purification Board, which now requires notification prior to treatment of fish with Nuvan, only became aware of the problem ten years after the use of the chemical was established in the industry.[25]

An equally toxic substance which was also used freely by fish farmers for a number of years is an antifouling agent tributyl tin (TBT). Nets were coated with this compound to discourage weed growth, and it was widely used in yacht paints during the 1970s. Evidence that minuscule residues of TBT were responsible for sex changes and reproductive failure in oysters, mussels and other shellfish led to a French ban on TBT paints in 1982.[26] In 1986, amid mounting concern, the Scottish Salmon Growers' Association placed its own ban on TBT preparations, and in 1987, following research which showed that TBT was being accumulated in the flesh of farmed salmon, the government eventually prohibited its use on cages and rafts. Even then, no advice was offered on safe disposal of existing stock, and one Shetland fish farmer was fined for dumping a 45 gallon drum of TBT over a local cliff.[27]

The scale of chemical pollution, if not its effects, can be assessed with reasonable accuracy. Much more difficult to estimate is the size of an iceberg whose tip is only intermittently visible, and about which most fish farmers are reluctant to talk: the deliberate and often illegal killing of wild marine mammals and birds which trespass on their cages. Not surprisingly, the presence of huge numbers of fish in a small area of water acts like a magnet to piscivorous birds and mammals. In recent surveys, about 80 per cent of fish farms claimed to suffer damage from seals, while around half were affected by herons, cormorants and shags.[28,29] The Department of Agriculture and Fisheries for Scotland (DAFs) is permitted to issue licences for shooting seals or birds to prevent serious damage to fisheries, but in the buccaneering spirit which characterizes the industry, it appears that the majority of sharpshooters prefer to operate outside the law. In 1987 DAFS issued thirty-six licences to shoot seals, and their statistics indicated thirty-six reported kills; and their current policy is not to issue any licences for shooting birds at marine fish farms.[30] A survey by Alison Ross for the Marine Conservation Society (MCS) in the same year produced the conflicting statistics that only 28 per cent of fish farmers held DAFS licences, though 72 per cent claimed to shoot seals, with 206 reported killings at the forty-seven farms she visited. The same proportion of fish farm operators shot birds, and estimated they killed 35 herons and 150 cormorants or shags in a year.[31] Extrapolating from these figures, the MCS claims an annual mortality of over 1000 seals and 200 herons.

Accurate numbers of predators killed at fish farms are impossible to obtain, but local knowledge supports the Marine Conservation Society figures against the official statistics. Shoot-outs of a dozen or more seals at a time are reported, and at one farm it is claimed that employees are rewarded for their efforts with a can of beer for each bird they shoot and a bottle of whisky for every seal. Elsewhere, dead seals have been anchored by stones to the bottom near cages, like the crows that bloody-minded farmers dangle from trees or fences as an example to their fellows. In a recent prosecution brought by the Royal Society for the Protection of Birds, a fish farm manager was convicted of killing eight herons using gin traps. When the birds were caught by the leg, they fell into the water and drowned.[32]

The industry's response to such horror stories is to emphasize the losses they suffer due to predation. Estimates vary from hundreds of thousands to millions of pounds' worth of damage.[33] However, it is very unlikely that shooting seals and birds is an effective way of significantly reducing these losses. The Nature Conservancy Council report that shooting programmes have very little impact on the number of piscivorous birds visiting sites where food is readily available; and while seals show more attachment to particular sites, their raids on fish cages are usually from deep under water on moonlit nights, when they are unlikely to be deterred by trigger-happy marksmen.[34] Furthermore, shooting tends to be indiscriminate, as only a minority of fish farm workers can differentiate between common and grey seals, and some even confuse otters with mink. Only a minority of fish farmers deliberately persecute otters, but they and other rare species are at risk from unintentional disturbance.

A better approach to the problem of predation would be to improve the design and standard of operation of anti-predator nets. At present, many farms use curtain nets open at the bottom to protect the cages against seals and diving birds. In strong currents these nets are sometimes pushed on to the inner bag net which forms the actual cage, allowing fish to be bitten by the predators. Better separation between outer and inner nets, and closing the outer net at the bottom to avoid seal attacks from underneath could reduce losses below water level. The existing designs of anti-predator nets allow many birds and seals to become entangled – indeed, some fish farmers set their nets deliberately in such a way as to act as traps – and the Nature Conservancy Council have made a number of recommendations as to how these mortalities could be avoided.[35] Above the water line, top nets should also be used to protect the stock from birds. Herons are adept at standing on such nets and fishing through the mesh, so again there is scope for improvement in design. Birds which fish from the air, such as gannets and ospreys, are not a major problem for fish farmers; though there are reports of gannets being trapped in cages unable to take off, and in Eastern Europe a number of ospreys have been killed by colliding with nets stretched above the water as they dived for fish.[36,37]

Fish farming is a young industry, hardly more than ten years old, and it is to be hoped that with experience and the establishment of more effective controls, many of the above adverse effects on the marine environment can be reduced. But more radical critics of the fish farming enterprise produce broader environmental arguments against it, which apply almost equally to many other forms of intensive livestock keeping. In opposition to the advocates of modern farming methods, who point to ever-increasing production as the basis for progress and the answer to world hunger, there is increasing criticism from people of broadly 'green' persuasion to the effect that low-tech conservationist agriculture offers the only long-term hope for the earth.

12

Farming for a Healthy Planet

When God made the world, he did so for the benefit of mankind. This belief, which can be traced back at least to Aristotle, is so convenient and agreeable that mankind has generally had no difficulty in accepting it without question. It is true that Lucretius asked sarcastically whether deserts and distempers were really created for man's sole pleasure, and that Porphyry pointed out that if one believes pigs were specifically made to be eaten by men, one might as well believe that men were specially made to be eaten by crocodiles; but on the whole people have been happy enough to regard themselves as the centre and final cause of God's creation. The elaboration of this belief could reach quite fanciful proportions: cattle and sheep had been given life so as to keep their meat fresh till man needed it, God had made the horse's excrement smell sweet because he knew that men would often be nearby, and so on.[1]

Such anthropocentric faith implied that nothing in nature was sacred on its own account, but all was merely a resource to be used by Man entirely as he felt fit. This is, in fact, still widely held in our own age; as the World Conservation Strategy states quite bluntly, 'Conservation, like development, is for people'.[2] This is fair enough, and seeing things from a human point of view is probably the only rational basis for debate; so long as it is accepted that people have moral and aesthetic priorities, as well as material ones. But the past hundred years or so have witnessed a radical change in the way we see our place in the world, and it is as well to remember that our priorities are not always the same as those of our ancestors.

The rise of science, with its discovery of countless new species, its demonstration that the solar system is only a tiny part of the galaxy, and its unravelling of the fossil record to chart a history millions of times longer than that recorded in the Bible, has inevitably changed our view of man's importance in the scheme of things. At the same time as our own cosmic significance has dwindled, the finite nature of our planet has become more apparent, and our capacity to damage it more alarming.

Influenced by these changes of perception, the Greens have followed the Romantic Movement, with its sublime sense of 'something far more deeply interfused', in trying to restore to Nature a value which has a kind of spiritual reality, and is not merely instrumental towards the achievement of human ends.

It is against this background of changing values that we have to consider the question of resources and exploitation. Does the fish farmer have our full approval in wresting the rewards of his labour from an alien environment, and dealing with nature's predators according to the law of the jungle, or is there a sense in which the fate of humanity is so linked with that of the rest of nature that it is impossible to damage it without at the same time damaging ourselves? If the past is anything to go by, we shall continue to deal ruthlessly with competing species by starving, shooting, trapping and burning them out with little regard for anything but our own immediate object. The entire agricultural landscape is a testament to such policies. The bear and the wolf, driven out of most of Europe, are threatened in North America by persecution and destruction of habitat. In Africa, fencing for cattle has cut off the migration routes of wildebeest so that tens of thousands have died from starvation and thirst. Throughout the world, habitats are being destroyed by slash-and-burn agriculture, and bird populations continue to be decimated by poisonous insecticides.

Eventually, though, a point must be reached where sheer self-interest should persuade the hardest nosed businessman to tread more lightly on the earth. We can never replace the whales, the birds, the plants we have pushed over the brink of extinction; and it is doubtful that we can ever reclaim the deserts we have made, or undo the changes we are wreaking to the world's climate. It is becoming increasingly apparent that there is no island of technological security from the threatening environmental chaos, and while the conversion of world leaders to Green policies may stem more from prudence than from compassion, it is none the less to be welcomed. The optimistic, expansionist dreams of the colonial and industrial eras are overdue for replacement by a more sober realization that we have only a limited amount to share around, and are likely to have even less in the future.

In a finite world, how can we make the most efficient use of the available land? Is it by cramming more and more livestock into ever-smaller spaces, creating 'cities' for farm animals as we have for human beings? That we should aim for efficient land use seems beyond doubt. Even in the days of 'Go West, young man', when the world was a much emptier place, every acre gained by the settler was an acre less for the Indians and their buffalo herds; an exchange which was justified (if at all) by the white man's claim to make better use of the land. And now we are more aware of the way in which successful farming tends to banish wildlife, it would seem there are good reasons for getting what we need from the minimum area of ground so that more habitats can be preserved in an unspoilt state.

The whole rationale of farming is to produce the food we need to stay alive, which for anyone not on a slimming diet means, as well as adequate protein and other essential nutrients, an available energy input of something around 2000 kcal per day. The ultimate source of this energy is the sun, and since animals are incapable of using the sun's rays to synthesize food directly, we rely on plants to perform the vital first step. On good soil and under favourable conditions most plants can make use of about 2 per cent of the energy which reaches the ground from the sun, and even with the most prolific crops less than half of this is available as food yield. Thus an acre of ground which receives about 2000 million kcal from the sun during the growing season might yield 4 tonnes of wheat, with a food value of 12 million kcal, or enough to feed sixteen people for a year. If instead of growing crops for direct human consumption, we use the land to grow fodder for herbivorous animals which are then slaughtered for meat, the productivity in terms of human food per acre drops by a factor of ten. And if anyone were rash enough to try farming carnivorous animals for meat – animals which would eat other animals that had themselves fed on plants – yields per acre would be around a hundredth of the usable food that could be obtained from plant staples.[3] If food value is assessed in terms of protein rather than energy, the results are essentially similar, with the protein from an acre of crops ranging from five to fifteen times that from the same acre devoted to meat production, depending on whether cereals, legumes or leafy vegetables are grown.[4]

So, however many broiler chicks can be crammed into a shed, they still need to be fed on arable crops essentially similar to those we could live on ourselves. Broilers have the highest rate of food conversion of any form of livestock, but it still takes more than 4 kilos of feed to fatten a 2 kilo broiler, less than half of which is usable meat. Arguments like this are often used to justify vegetarianism as a far more efficient way of feeding the world's population. In most countries, in fact, people eat mainly vegetable crops already, and meat is regarded as the luxury it undoubtedly is. Yet even today, there are people who still believe that by exporting factory farm technology to Third World countries we are somehow helping to overcome their food shortages.

In Britain and the United States, more than half the cereal crop is fed to farm animals. If we ignore the damage to the environment caused by high input cereal monoculture, and the damage to our health caused by too much flesh in the national diet, raising all these meat animals might be regarded as a luxury the affluent West can still afford. But in Europe particularly, the crops we can grow ourselves are not on their own the most cost-effective ingredients for animal feedstuffs. As a result the EC imports something like 14 million tonnes of linseed, soya and other crops for animal feeds from countries in the Third World which often cannot feed their own human population. In 1984 Britain's imports of such crops from Ethiopia were worth more than £1,500,000 – at a time when that country was in the grip of the most appalling famine.[5] While poor countries undoubtedly need the income from cash crops, it is a tragedy and an obscenity when their agriculture can feed our farm animals but not their own children.

The craving for a more Western life style is leading to an increase in meat consumption in many countries. The Japanese are now importing two and a half times as much beef as they did ten years ago.[6] This they can doubtless afford, but in the major cattle producing country of Mexico, 80 per cent of the population is claimed to be undernourished, while most of the available food supplies are fed to livestock.[7] The manufacturers of cages, equipment and drugs have formed consortia to sell the technology of factory farming overseas. In 1985 it was reported that Pakistan intended to double poultry meat production over the next five years; at a time when the poultry industry was already facing a shortage of grain and demanding special privileges over imported feeds. The Libyans have bought an integrated broiler project with a capacity of 5 million birds a year, while in Egypt, Iraq, Jordan and Zanzibar similar complexes are already in operation.[8]

High protein livestock feeds do not just use plant ingredients. Fish meal is an important part of poultry food, and a vital constituent of the diet of farmed fish. Fish farming already accounts for 5 per cent of the 300,000 tonnes of fish meal consumed annually in the United Kingdom, and at its present rate of expansion could soon use up to 15 per cent.[9] The fish meal industry started out as a way of using surplus catches for which no market could be found; a hundred years ago in Scotland, when the supply of herring exceeded the demand from the local curing yards, the remainder would be rendered into meal and oil. This carries on to a limited extent, with offal rejected during the processing of fish providing an additional source of supply; but most of the world production of fish meal is now based on industrial fisheries, whose catch is intended for reduction to meal from the start. Such fisheries have been criticized in recent years for their effect on the supply of fish available for human consumption, and for their effect on other wildlife. Between

1965 and 1970, Norwegian purse seine boats were catching up to 600,000 tonnes per year of herring for fish meal in Scotland waters alone. As a result, the herring stocks of the North Sea collapsed totally, and have not yet recovered. More recently, the local sand eel fishery in Shetland has been blamed for repeated serious breeding failures at the important sea bird colonies on the islands. About 40 per cent of the British population of Arctic terns nest in the area, 61 per cent of Arctic skuas and about 20 per cent of puffins. Since 1983 virtually no terns have bred successfully, and recent studies have shown an extraordinary increase in skua chick mortality, and a total failure of puffins to breed in the past few years.[10] These species all rely heavily on sand eels to feed their chicks, and a dramatic crash in their total numbers is certain to follow these disastrous breeding performances.

Sand eels are not eaten by humans, nor are a number of other species that are caught for fish meal, but they are part of a food chain that leads up to fish more suitable for human consumption. Industrial fisheries often involve unacceptable by-catches of more valuable species; for example, Norway pout cannot be caught in the North Sea without also catching large numbers of immature haddock and whiting. And the feeding of such edible species to farm animals is very wasteful: the 20,000 tonnes of farmed fish produced in Britain in 1986 consumed pelleted feed incorporating 20,000 tonnes of fish meal derived from 100,000 tonnes of raw fish.[11]

Partly as a result of the demands of factory farming, the amount of fish caught in developing countries and shipped to developed countries is increasing. About a third of the world's fish catch is now fed to farm animals.[12] As in other fields, the poorer countries have to run faster to stay in the same place. In the ten years to 1983, per capita fish consumption in developed nations went up from 23 kg per year to 27 kg, with fish prices rising faster than meat prices, while consumption in developing countries stagnated at less than one-third of this level.[13] In a heart-rending report from Peru, Jennifer Amery described the fishing port of Chimbote, where 99 per cent of the catch is exported for pig or poultry food, yet two-thirds of the population is starving, and children are dying for lack of protein.[14]

Famine is not entirely a matter of lack of food resources. Even with all the waste involved in our factory farms, world food production in 1986 exceeded demand by 10 per cent, yet 40,000 children died daily from hunger and related causes.[15] Nor is there any ultimate solution – whether we subsist on hamburgers, boiled rice or protein synthesized from algae – which does not involve getting away from unlimited population growth. But however you look at it, the contribution of intensive livestock keeping to ecologically efficient, nutritionally adequate global agriculture is entirely negative.

Some vegetarians would go further, and argue that all livestock farming is tarred with the same brush. If beef cattle grazing a field can only provide a tenth of the food that could be grown as crops on the same area, what justification is there for keeping them? Even the dairy cow only converts about a quarter of her protein intake into milk. But not all pasture land is fit for growing crops that we would want to eat; many cattle and most sheep are kept on land which is too wet, too dry, too steep or otherwise insufficiently fertile to produce worthwhile yields of food crops. Even intensively fed dairy cattle get only about 15 per cent of their energy requirements from cereals such as barley or wheat, the rest of their diet comprising materials unsuitable for humans, such as the fibrous residues from soya or groundnut oil production.[16] All grain crops produce straw, which can be fed to cattle as a useful source of energy. And pigs and poultry were traditionally kept on scraps, as they still can be in small-scale rural agriculture. In the 1960s as many pigs were being kept on household scraps in China as were being raised on grain in the United States.[17] Colin Tudge points out that even feeding grain to livestock can be justified if it is done in moderation. To insure against bad years, farmers must always aim for a surplus of staple crops, and when this is achieved nothing is lost by feeding it to livestock. In good years the luxury of meat eating can be indulged more freely; in very bad years the animals provide a reserve food stock which can be slaughtered in time of need.[18]

Modern intensive farming is obviously not the most efficient way of utilizing the available cropland area of the globe, and the energy that falls on it; but its efficiency looks worse still when we consider the other inputs it requires. Unlike the traditional mixed farm, which was virtually self-sufficient, and really did derive almost everything from sun and rain, including the crops to feed horses which helped with all the hard work, the modern farm has an insatiable appetite for diesel oil, electricity, synthetic fertilizers and planned-obsolescent hardware. When all these inputs are expressed in terms of the fossil fuels required for their manufacture, it has been calculated that in the United States agricultural energy efficiency has declined ten-fold since 1910, to the point where 10 calories of energy are now consumed for every one calorie of food produced.[19] British figures for a breakdown of the national farm energy budget show that about a third is used to heat buildings and power machinery, another third in the production of fertilizers, and the rest goes into feeding stuffs, machinery and other agrochemicals.[20] Still more energy is expended in processing, transporting, retailing and even cooking produce, after it has left the farm gate.

How much should these figures worry us? It would be rash to claim that we should only grow food crops like potatoes, which provide more calories than they consume, and should eschew at all costs the profligate tomato which gives but one calorie for every hundred used to grow it; but the inefficient use of fossil fuels in modern agriculture introduces a whole range of questions about its sustainability.

Synthetic and natural mineral fertilizers alike are based on non-renewable resources: however vast reserves seem at present, they may ultimately be used up, and we shall be forced to look for alternatives or do without. They are a theoretically more difficult, but actually less pressing problem than some of our renewable resources, which will carry on for ever if we are careful, but which we can very soon destroy altogether if we are not.[21] Mercifully we have learnt how to do without the oil and meat of whales, so if by good fortune they are saved it will not be for margarine and candles; but if we hunt to extinction the herring, the anchovy or the tunny fish, we may look back in hunger as well as with regret.

Many of the most pressing conservation problems of the age are closely connected with changes in agriculture. The trend towards specialist and intensive livestock farms is part of a general movement to larger and more specialized agricultural holdings; and their establishment is one of the principal pressures making traditional mixed farming uneconomic. Unfortunately, what is economically favoured in the short term may be quite disastrous over a longer time scale. Unlike those who actually live on and work their home farmlands from generation to generation, the big businesses and banks which dominate the modern agricultural scene have little concern for the long-term sustainability of their projects. If it is profitable now to invest in growing wheat on the fenlands, who cares if tomorrow the fields are a poisoned desert? Capital is mobile, and as the past and present agonies of the American agricultural scene make abundantly clear, money lenders do not hang around when they are no longer making a return.

The ultimate renewable resource of agriculture is the soil itself. In the United States erosion of croplands, which are mainly used to grow animal fodder, is taking place at a rate of at least two billion tons of soil a year; and in 1970 the National Academy of Sciences reported that the nation had lost a third of its topsoil.[22] World wide, it has been estimated that a third of all arable land will be desert by the turn of the century. The situation with grazing land is equally bad, as more and more cattle are raised in fundamentally unsuitable areas to provide beef for the fast-food industry. In the Sudan alone, over half a million acres a year are being turned into desert by cattle grazing.[23] The subdivision of grazing land into small enclosures may make life easier for the farmer, but fences also block the movement of wildlife and result in the starving out of indigenous species. Even when permanent pasture is physically stable, the use of fertilizers to sustain higher stocking rates can cause deterioration in the quality of the herbage. Magnesium and copper deficiencies may result, and increased stocking densities inevitably produce a greater risk of infection by intestinal parasites.[24]

The preservation of soil fertility through traditional mixed farming methods is the main objective of the organic farming movement, and in Britain the Soil Association distributes numerous tracts on how this should be done. Not that the degradation of agricultural land is a new phenomenon – the regression from forest to desert was common knowledge more than two hundred years ago:

> in woods that had been preserved for a long time without being touched, the layer of soil that serves for vegetation would considerably increase; but as animals return less to the earth than they draw from it, and men consume enormous quantities of timber and plants for fires and other purposes, it follows that the layer of vegetative soil in an inhabited country must always diminish and become in the end like the terrain of Arabia Petraea, and so many other provinces of the Orient, which is in fact the area of most ancient habitation, where today we find only salt and sand.[25]

Today, rather than follow the path to agricultural nemesis described by Buffon, we have learnt to accelerate the process: without the timber being used at all, tens of millions of acres of tropical forest are simply burnt down every year to make way for grazing cattle. This may be economic sense, but it is ecological insanity. Not only is the pasture produced often unstable, being rapidly degraded by erosion in the space of very few years, but the destruction of the rain forests results in tremendous loss of species, and may actually threaten the world's climate.

Even in regions which are ideally suited to grazing herbivores, cattle may be quite the wrong species to introduce. In Africa, the environmental damage wrought by cattle has been appalling. They require more water, and are far more destructive of the native vegetation than indigenous species such as antelope and gazelles. Ranching wildebeest or other native ungulates could be up to ten times more productive per acre than inappropriate European style cattle operations.[26] At present native species are persecuted by cattle ranchers because they compete for grazing, and also carry a number of diseases transmissible to domestic stock. One problem with exploitation of wild animals for meat is the prevalence in wild stock of diseases transmissible to humans, such as tuberculosis, brucellosis and trichinosis. In the Serengeti most herbivores are so infested with muscle parasites that canning and sterilization is recommended for all game meat from the area. South African springbok, on the other hand, are remarkably free from transmissible diseases and parasites.[27]

While there is much the farmer can do on a local scale to improve or destroy his soil, it would until very recently have seemed impossible that his actions could have any influence on the weather. Yet the alarming phenomenon of global warming which has recently achieved such prominence is not just the result of burning fossil fuels, or using aerosol sprays, but to a surprising extent is the result of cumulative changes in agriculture and land use. Deforestation contributes carbon dioxide to the extent of about 20 per cent of the total greenhouse gases we are currently adding to the atmosphere, and agriculturally produced methane has almost as much effect.[28] The latter is produced by paddy cultivation of rice, and by the gut bacteria of ruminant animals. In addition, large-scale tropical deforestation could affect the worldwide atmospheric circulation in such a way as to change the weather patterns experienced in temperate latitudes.[29] Similar remote effects have been predicted as a result of desertification in other parts of the globe.

In naturally dry regions, the agricultural demand for water is enormous. The United States uses a staggering 11,000 litres (2500 gallons) per head of population per day to irrigate crops grown for livestock and as drinking water for farm animals. Consequently the huge Ogallala Aquifer below the Great Plains grain belt is suffering serious depletion, and its quality is threatened by pollution from fertilizers and insecticides.[30]

The ecological effects of small concentrations of chemical residues are notoriously difficult to monitor. It took twenty years for the impact of DDT on the global environment to be detected, and we still know far too little about the mechanism of biomagnification whereby, for example, a concentration of dieldrin in the soil (in parts per million) of 0.06 can become 1.5 in earthworms and nearly 15 in garter snakes.[31]

Only time will tell what has emerged from the Pandora's box of genetic engineering. The European Community have agreed on a directive on deliberate release of genetically engineered organisms into the environment, and in Britain the Health and Safety Executive has produced a masterfully obscure set of Guidelines; but as nobody has the slightest idea of what to expect if things go wrong, such governmental flannelling is of very doubtful value. In the Federal German Republic, the government has at least proposed fines of up to £33,000 if things go wrong, though this may not frighten the average multinational a lot; while the Confederation of British Industry has come up with the dubiously helpful advice that industry should be cautious over allowing more public access to information on genetic engineering developments.

Whether we are reassured or alarmed by the public relations campaigns of the multinational corporations, it is increasingly apparent that we are all going to be exposed to the environmental consequences of present agricultural policies. Some of us are lucky enough, and well enough off, to be choosy about what we eat, but we all have to drink the water, breathe the air and endure the climate. We are at last beginning to realize that pollution knows no boundaries. Banning toxic chemicals at home is not much use if they come back to us in imported fruit, or are carried round the world's oceans, or are wafted up to destroy the ozone layer. Destruction of species and poisoning the environment cannot simply be national issues. And insofar as modern agriculture contributes to these global problems, with its emphasis on the quick commercial return and its lack of concern for sustainability, it presents us with an urgent challenge far nearer home than the Amazonian forests or the cattle ranches of Botswana. By taking care over what we buy, and by putting our own agricultural house in order, we can all do a little for the future in the spirit of the old countryman's saying: 'We don't inherit the earth from our ancestors, we borrow it from our children'.

Part V
DO WE GET WHAT WE WANT?

13
Politics and Profits

Food is a political subject in any country which supports an urban population. Once people must depend on growers and merchants, the quality and sufficiency of food supplies are matters of concern for a government which wishes to remain effective. Politicians cannot control the weather, or turn a bad harvest into a good one, but since the Pharaohs of ancient Egypt prudent leaders have taken steps to protect their populace from the vagaries of climate, and from extremes of exploitation by middle men. Profiteering by merchants who achieve monopolies was doubtless widespread before the days of Diogenes, and is certainly still with us, but if unchecked it is a sure contribution to political instability.

For nation states which develop to the point where home grown produce can no longer satisfy ever-increasing numbers of city dwellers, guaranteeing food supplies becomes a matter of strategic importance in foreign policy. The prosperity of ancient Rome was dependent on imports of grain from North Africa; and the industrializing nations of modern Europe found their colonies invaluable both as markets for manufactured goods and as sources of cheap food. The growth of international trade in crops, livestock and other food products means that home prices do not simply depend on local preference and abundance, but are influenced by worldwide gluts and shortages, and by every success and set-back in foreign diplomacy. So it is impossible to make any sense of national agricultural policy in terms of a relation between farmers and consumers without reference to the current international policies and priorities of the government. Nor can it be decided in isolation which methods of livestock production make economic sense, and whether factory farming is inevitably more profitable than traditional husbandry. Before such questions can be addressed, the various effects of governmental intervention must be considered.

Cheap food from countries overseas with favourable climates and lower labour costs seems like an excellent idea in peacetime, but the risk of blockade should war break out has traditionally discouraged prudent statesmen from allowing a complete run down of home agriculture. Foreign competition is discriminated against by customs barriers or state subsidies to home farmers, or both. Even where there is no threat of shortages, tariffs on food products are often used as bargaining points in negotiations over trade in other commodities; or trade may be stopped completely for strategic reasons, as on the occasions when the United States has embargoed grain exports to Russia.

British policy on imported foodstuffs has varied considerably since the beginning of the last century. Following the Napoleonic Wars, imports of wheat were subjected to duty to protect home producers, but the resulting increase in the price of bread caused great hardship among poorer people, and the Corn Laws were finally repealed in 1846 during the Irish famine. A long period of free trade followed, during which we bought produce from wherever it could be produced cheaply: grain from the United States, beef from the Argentine, butter and lamb from New Zealand and so on. Our home farms continued to produce these foods, with increasing emphasis on livestock farming, where a premium could be demanded for fresh home produced meat and dairy products. As well as being cheap, this policy protected us from the risk of shortage should crops fail in any particular supplying country, or at home. The depression of 1929 led to adoption of protectionism against a wide range of imports, including agricultural produce. In the Second World War, the importance of a viable home farming industry was underlined as Hitler's U-boats threatened the freedom of the seas which Britain had so long taken for granted, and in the post-war years we have continued digging for victory with scarcely diminished enthusiasm; increasing year by year the proportion of home produced food, and cutting down trade with our traditional suppliers. But in the age of the atom bomb, is it likely that we shall ever need to cope with a siege of several years' duration? If not, the military value of agricultural independence is surely illusory; nevertheless, fifty years after the Battle of Britain, the rallying cry of self-sufficiency in food staples is still to be heard, and it is a brave person who dares question the sense of pursuing such a policy.

Since our entry into the European Community, British farmers have had slightly more exposure to competition from their European counterparts, but as far as the rest of the world is concerned the European Community has adopted a fiercely protectionist stance. In 1983, beef was subject to an import levy of £2.10 per kilo, butter of £1.35 per kilo and sugar 22 pence per kilo.[1] Most people still remember the leap in food prices which accompanied our joining the European

Community; and the overall cost to the consumer of EC food policies is about £10 per week for every family. Three-quarters of all farm support goes to the top 25 per cent of farmers: £5000 per year for each of them.[2] The real purpose of agricultural protectionism is quite simply the same as in any other industry: to shelter the home producer from unwelcome foreign competition. Whether farmers are protected by import barriers or by direct subsidy, their protection costs the rest of us. Import levies push up the price of food, while governmental payments to farmers have to be funded from general taxation. Where, it may be asked, are the social benefits to make these costs worth while?

One benefit, eagerly anticipated by the Treaty of Rome, was 'a fair standard of living for the agricultural community'. But farm workers are still among the lowest paid in the country, despite the fact that fewer than a third of the number of people directly employed in agriculture in 1946 now produce a much higher output.[3] This apparently huge increase in productivity has not made the farming enterprise more profitable; while output has risen, the cost of fertilizers and depreciation of expensive machinery has risen even faster, so that net farm incomes have actually declined in real terms. At the same time the amount of public money spent on agriculture has increased enormously. By the 1980s, the cost to the taxpayer of farm price support has been estimated at more than twice the total net income of all British farmers (see Figure 5).[4] Doubling the amount (in real terms) spent on price support between 1970 and 1980 did not stop average farming incomes falling by 50 per cent over the same period.[5] Quite obviously, if the object of the exercise is simply to sustain the workers on the land, the present method of subsidy is grossly inefficient.[6]

What, then, happens to the immense amounts of taxpayers' money that enters the maw of the Common Agricultural Policy (CAP)? It is not distributed to reward the labours of those who grow the crops, nor, contrary to popular belief, is it all lost in the bureaucratic labyrinths of the European Commission, or used to bribe the Russians to relieve us of mountains of rancid butter. It is true that some funds are lost in this way, and that £2750 million per year – about 10 per cent of the entire community budget – is siphoned away by frauds, largely to the benefit of the Mafia; but if this were the whole story the CAP would have even fewer friends than it does now.[7] The safety net of the European Community's regulated price system has done far more to encourage capital investment in farming, than to improve the rewards for those who actually labour in the industry. Because a market price is guaranteed, whatever the extent of overproduction, banks have been only too happy to lend for the purchase of fertilizers, machinery or additional land. The

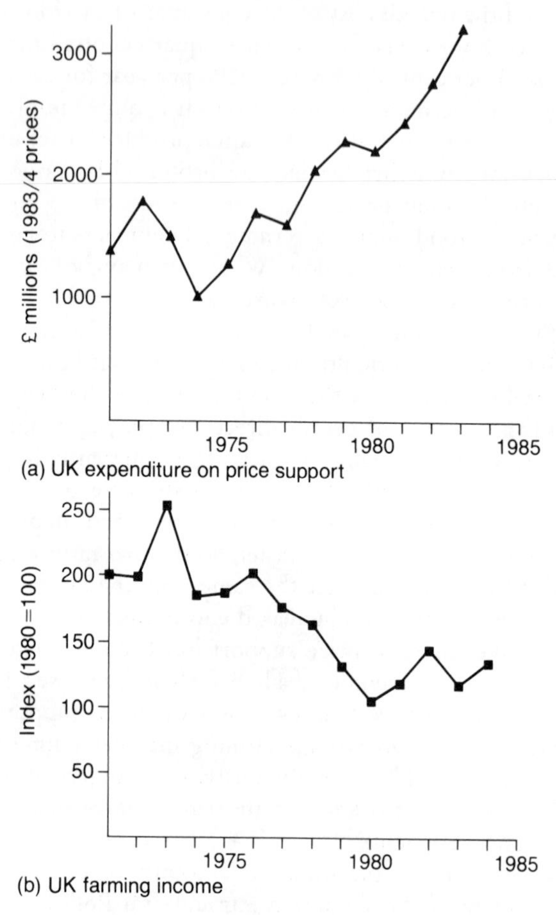

Figure 5 (a) UK expenditure on price support.
(b) UK farming income.

chemical industry has had a field day, with fertilizer use continuing to escalate despite soaring prices; little wonder that ICI were the most enthusiastic of all industrial campaigners for British membership of the European Community. Other manufacturers have also benefited, but the happiest people of all have been the land owners, as much of the public money involved has been capitalized in soaring land values.[8] This has been greatly appreciated by farm owners wishing to sell, but not by tenants who have had to pay higher rents, nor by those having to borrow capital to buy farms at inflated land prices.

Protectionism costs the taxpayer, but is good news for cereal farmers who are insulated from fluctuating world prices, for manufacturers, for land owners and for money lenders. And what is good for these worthy classes must be good for Britain, or so their lobbyists must convince the politicians. Unlike any other industry, agriculture has its own Cabinet minister who can be relied on to put the farmers' case – or at least the case of the wealthy land owning farmers who have sufficient spare time and influence to reach important positions in the National Farmers' Union (NFU). The aims and methods of the NFU lobby have been clearly described in *Farmers Weekly*: '[It] means knowing who the most powerful people are on a certain issue and bringing them round to the farmers' point of view. It is the kind of lobbying which goes on at lunchtimes, in bars and at private dinner parties where Cabinet Ministers are mellowed by good claret and port. It's expensive and exclusive, but it works.'[9]

The Common Agricultural Policy effectively protects European farmers from the ups and downs of international agricultural trade, but at the cost of a pronounced destabilizing effect on world markets. The dumping of surpluses depresses world prices, to the detriment of other agricultural exporters. Abandoning our traditional trade with Australia, Canada, and Argentina has severely damaged the economies of these countries, and has lost substantial markets for British exporters of manufactured goods. Even worse is the effect of the CAP on the poorest countries of the world whose only possible exports are agricultural products, which can only be sold at a loss in competition with produce dumped by the European Community. Cane sugar growing countries such as Guyana and Barbados have been particularly badly hit, but their limited access to good port and the ears of Cabinet ministers has ensured that the devastation of their economies by the CAP has gone largely unnoticed.

The economics of intensive livestock farming compared with non-intensive methods depend partly on intrinsic factors such as feed conversion and labour costs, but are also subject to various aspects of government policy which discriminate against one type of system or the other. The inflation of land values by the CAP inspired philosophy of production at any price has encouraged farmers who wish to expand to make more intensive use of the land they already have, rather than buy or rent more costly acres. Intensive, silage-based dairy operations are thus favoured compared with keeping the cattle on open pasture, while large pig and poultry units can be built on very little land indeed if the animals are intensively housed. The British tax allowance system further favours investment in plant and buildings rather than in the purchase of land. A factory pig farm makes use of land worth perhaps £10,000 which must be purchased from the owner's capital, but the buildings and equipment costing perhaps £200,000 can be written off against income tax. Richard Body claims that no-one in his right mind would put this much of his own cash in such a venture, nor would any bank manager lend it to him; but the tax advantages to a large scale arable farmer or farming company make the intensive route much more attractive than the alternative of keeping the same number of pigs on land costing £200,000 in simple sow-huts costing £10,000 or less.[10] The orientation of agricultural grants and subsidies towards increased output at all costs inevitably favours intensification. The synthetic hormone BST may indeed produce a 40 per cent gain in milk output which will more than cover the cost of the new technological treatment, as long as milk prices do not falter. But with price control the result would be more and more gallons of milk poured down the drain, and with milk quotas more and more productive pasture 'set aside'. In an uncontrolled market, the marginal value of such expensive ways of increasing production might be very different. Finally, an aspect of the Common Agricultural Policy which acts against intensive livestock keeping is the inflated price of cereal fodder. The effect of import levies cutting off cheap grain supplies can practically double feed prices for pig and poultry farmers. The smaller scale free range farmer may save a little by feeding grass or scraps, but discounts on feed available to larger farmers, and the possibility of better feed conversion under intensive conditions could easily offset this advantage. Cattle farmers have reduced the impact of higher grain prices by locating alternative feed ingredients such as manioc and cassava, against which the European Community has not raised any import barriers.

The distorting mirror of agricultural policy makes meaningful comparisons between different husbandry systems difficult. As the discussion above shows, it is by no means clear what the criterion of an 'efficient' farmer should be. Is efficiency measured in terms of output per man, output per acre or profitability? If any of these, no dairy farmer could ever be efficient in the way of the East Anglian prairie barons, sucking their millions from the public coffers. Or is an efficient farmer one who makes a living from his land without being a drain on the taxpayer? In this case, efficient farmers are a dying breed. Body cites the case of the owner of 60 acres of poor pastureland whose pig-keeping enterprise became unprofitable when his import levy bill reached £30,000. With the aid of drainage grants he converted the ground to arable land, and is now paid £15,000 per annum to grow a poor crop of wheat. The farmer remained solvent, but the exchequer is worse off to the tune of £45,000 per year.[11] The economist E. F. Schumacher made the radical suggestion that the management of farmland and livestock should be directed towards the three goals of health, beauty and permanence.[12] But while all these attributes may have value for you or me, they are not yet widely appreciated by politicians or their worldly-wise advisors.

The economic advantages claimed for large scale livestock farming include reduced overheads per animal, the possibility of bulk discounts on feed, and the ability to meet the demands of major purchasers such as supermarkets without needing to sell through middle-men. Intensive livestock operations offer the further benefits of reduced labour costs, though at the expense of higher capital investment. These facts are indisputable, but there is considerably more doubt about what sort of system is more efficient in terms of feed conversion or animal health. Comparisons are difficult, because during the period of general change-over to more intensive methods there have been substantial advances in animal breeding and feed formulation; also intensive systems have often been introduced by the more commercially acute farmers whose management tends to be tighter, giving a more favourable view of the systems' efficiency.

Regardless of husbandry methods, flock or herd size can affect the performance of animals. There is evidence that broilers put on weight more rapidly when kept in small groups: birds kept in a flock of twenty were 50 per cent heavier after nine weeks than similar fowls in a group of 30,000.[13] Likewise, in an experiment to compare the performance of Friesian steers kept in groups of eighteen or twenty-eight, the former gained weight more rapidly.[14] Similar effects have been observed with pigs, but while weight gain is more rapid in small groups, at least one study has shown that feed conversion is more efficient when larger numbers are kept together. It is very difficult to keep all the other variables constant in studies of this sort. Some of the loss of performance in larger groups can probably be put down to feeding difficulties caused by the design of the livestock housing. In crowded conditions, the more timid animals fail to get their rations, and therefore grow more slowly.[15]

Battery egg production is generally agreed to be considerably cheaper than free range, mainly because of the larger number of eggs produced (see page 28). Shop prices at present exaggerate the difference, partly because free range eggs are in short supply, and partly because consumers will happily pay extra for large free range eggs, but not for the undersized ones produced when the hens first start to lay. Paul Carnell has estimated production costs of 44 pence per dozen for battery eggs at

1980 prices, compared with 67 pence for free range; a more recent estimate, which may not be strictly comparable, gives figures of 37 pence and 58 pence.[16,17] At present, free range eggs command a substantial premium, but if battery cages were abolished, retail prices would probably increase by around 10 per cent for deep litter or perchery eggs, and by up to a third for free range.[18]

The cost advantage of indoor versus outdoor pig herds is less clear cut, though the vast majority of large pig operations are housed inside. As land prices slip back from the record levels of the early 1980s, outdoor pigs are beginning to look more attractive again, particularly for a farmer starting out on little capital. Sows at pasture get some of their food requirements by grazing, but this is offset by extra energy needed to keep warm in winter. Data from the Meat and Livestock Commission show that feed consumption and breeding results are very similar for indoor and outdoor herds.[19] Under cover, figures from the National Agricultural Centre's pig unit suggest that the use of electronic sow feeders reduces the annual feed cost per sow by nearly 10 per cent compared with the more old-fashioned dry-sow stalls or straw yard.[20] In his review of the economics of intensive systems, Carnell throws doubt on the extent of savings achieved by early weaning, and the use of farrowing crates, both widespread practices among pig farmers that have been criticized on welfare grounds.[21]

Efficient feed conversion is obviously a major concern of any farmer who wishes to remain competitive, but the less precisely quantifiable variable of animal health is equally important. Although health problems which do not affect production are often scandalously ignored in battery farms, there comes a stage at which economic performance is bound to suffer. Epidemic disease can wreck profits from the most theoretically efficient flock, and mortality rates in growing and breeding stock must be kept to a minimum if the overall productivity of the unit is to be maintained. Nowhere is this more true than in dairying, where as a general rule the farmers who make most money have the healthiest cows. A recent survey of dairy farms showed very little difference in milk output or feeding costs between the most profitable 25 per cent of farms and the least profitable 25 per cent. The major difference in physical performance was that replacement rates were lower in the more profitable herds, presumably because fewer cows were having to be culled for infertility, mastitis, lameness and poor performance.[22] Economic performance was only tenuously linked to the productive efficiency per cow, and much more influenced by fixed costs such as wages and overdraft payments, but none the less the most successful farmers did appear to have the healthiest cows.

Neither sheer size nor the use of the most technologically advanced methods is a guarantee of profitability. The small farm is still often a family business, and can provide a higher input of unpaid labour; while very large operations are usually funded with borrowed money, and therefore more vulnerable to the erosion of profits by interest payments. Statistics repeatedly show that output per acre declines with increasing farm size, even though larger farms still make bigger profits. Because so many farmers are running their own businesses, it can be expected that within the limits set by the need to remain profitable, there is scope for some subjective judgement about what sort of a business they wish to run. In the post-war years, modernization was both fashionable and profitable, but more recently the demand for organic foods and free range eggs has enabled a number of farmers to become highly successful supplying these markets. If environmental awareness and concern for animal welfare continue to grow, less intensive farming will inevitably be associated with greater job satisfaction, and become more popular.

From the farmer's point of view, intensive methods tend to mean much worse conditions of work. It is true that he spends more time under cover, and does not have to brave snow and wind, but few people would claim that job satisfaction or health have benefited. The atmosphere in high density animal housing is not a healthy working environment, but up to a million American farmers now work in such buildings. A study of pig farm workers in Iowa revealed that as many as 70 per cent suffer from respiratory difficulties; and another American investigation discovered hazardous levels of carbon monoxide, ammonia and hydrogen sulphide in pig houses, as well as high concentrations of air-borne bacteria and fungally contaminated dust particles.[23,24]

The attractions of a particular style of animal husbandry to the farmer are not simply its intrinsic profitability. Farmers have to make a living, but each has his particular skills and preferences, as well as the opportunities of his individual situation. Most of the economic factors influencing his choice are highly sensitive to governmental policies, as witnessed by the frantic changes between different types of farming which characterized the years following Britain's admission to the European Community. As well as the politics of international trade, more local issues such as regulations regarding animal welfare or environmental pollution can affect the viability of particular methods of farming. But while our trading policies, and those of the other European countries, have been developed very much by, with and for the farmers, the limitations imposed on agriculture in the name of public health, pollution control or animal welfare have invariably originated outside the industry altogether.

14
Serving the Consumer?

Farmers and food manufacturers foster a cosy public image of themselves as the generous providers of food to a hungry nation, the custodians of the countryside, and the guardians of all that is best about the British way of life. This pleasant fiction, recounted on billions of food wrappers and bolstered by food advertising now worth over £500 million per year, went largely unchallenged through the sixties and seventies. Following the relative austerity of the immediate post-war years, the dazzling choice of goodies in the new supermarkets was enough to convince most of us that we had never had it so good. Only very recently, with increasing concern regarding diet-related diseases, and a series of health scares over infected foodstuffs, has public complacency begun to be disturbed.

It is true that the majority of people in this country have a greater choice of diet than ever before, but there is mounting evidence that this freedom of choice is used wisely by only a minority of consumers. This is partly because we are by no means clear what the ideal diet is; and our appetites are easily beguiled by rich and sweet foods however great the evidence that the diet of poverty might be healthier. But the choice of diet is made far harder by the extraordinary lengths to which food manufacturers will go to deceive us. It will always be in the supplier's interest to dress up poor quality food as something better, and modern technology has a brilliant repertoire of tricks to achieve this illusion. Geoffrey Cannon describes a brand of oxtail soup, containing no fewer than eighteen additives and precious little oxtail, as 'a technological version of the gruel served up in Victorian workhouses and Soviet prison camps'. What is even more remarkable, when the oxtail and vegetable powders are replaced by fat-reduced cocoa and egg, a few subtle adjustments to the range of additives are all that is needed to change this doubtful brew into alluring 'fondant fancies'.[1]

Processed food is clever, convenient and vigorously promoted, but not necessarily wholesome or nutritious. Nevertheless, it accounts for

80 per cent of our national spending on food.[2] The results of a government survey on teenage eating habits were so appalling that they were suppressed for nearly three years before being leaked. Chips, crisps and cakes made up a huge proportion of the diet of young people, both at home and in school, since the obligation of local authorities to provide school meals conforming to prescribed nutritional standards was abolished.[3]

Even if we do tend to eat too much sugar and fat, surely the national diet now is vastly better than the meagre fare enjoyed by our ancestors? This is certainly what the food industry and the Ministry of Agriculture in Britain would like us to believe, and to give them due credit for it. Certainly in Victorian cities, where white bread and jam were the staple diet of many of the poor, malnutrition was rife. In the country, on the other hand, the diet of agricultural labourers was in many ways superior to what most of us eat today. Compared with the average modern British diet, the rural Welsh of 125 years ago consumed more protein, less fat, and more vitamins and minerals; furthermore, laboratory mice fed on this rustic diet lived longer than those given its present day replacement.[4] The health preserving role of vitamins was discovered in the period between the world wars, and gave a scientific basis to nutritionists' claims for the value of good fresh food, and to their campaigns for improvement in the national diet. To feed livestock on a diet known to be inadequate for maintaining health and fitness would be absurd, argued the prominent nutritionist John Boyd Orr, yet this was exactly what we were doing to children in the poorer half of the population.[5] In the 1980s rickets is no longer a problem, and the poor are generally much better off than fifty years ago, but consumption of vitamin C, iron and calcium has declined.[6] There is some dispute over what the recommended daily allowances should be, but it is little comfort that British standards are the worst in Europe, and far below those accepted by the United States and the USSR.[7]

The industrialization of farming can only encourage the trend away from fresh food towards more and more highly processed products. It is easier for the demands of the highly industrialized processors to be met by a more centralized and mechanized supply sector, producing crops or livestock of specified size and quality; if broiler chickens or intensively reared pigs are flabby and tasteless, the processing industry is quick to the rescue with chemical recipes to transform poor quality ingredients into crispy spiced Tandoori ready meals and Olde English style 'joints' of re-formed ham. But quite apart from what is added during processing, the loss of vital nutrients which were present in fresh food is a source of increasing concern. One of these is folic acid, a vitamin of the B group for which the government set recommended daily allowances in 1979, and quietly withdrew them again in 1981 when it became apparent that national average consumption was well below the recommended figures. Folic acid deficiency is well known to produce birth defects in animals, and there is now considerable evidence relating the incidence of spina bifida to lack of this vitamin in the pregnant mother's diet. Folic acid supplements are widely prescribed for pregnant mothers with any family history of spina bifida or anencephaly, and the makers of breakfast cereals would doubtless be happy enough to add synthetic folic acid to the other vitamin additives they already use. Such high-tech fixes are a poor substitute for a diet which is fresh and wholesome to start with.

The irradiation of food is a further threat to vitamin levels, as irradiation immediately destroys between 20 and 80 per cent of vitamins A, C and E in fresh foodstuffs, with further accelerated losses following treatment.[8] If, as some experts believe, these vitamins protect against cancers and are already at a disturbingly low level in the national diet, food irradiation poses serious long-term risks. Animals fed exclusively on irradiated foods have decreased fertility, lower growth rates, increased incidence of tumours and less resistance to disease. Some if not all of these effects are thought to be due to multiple vitamin deficiency diseases such as scurvy.

Health hazards from the products of factory farming, whether processed or not, have been discussed at some length in chapter 5. While it is inevitable that some of the 'scares' about food are based on circumstantial evidence, and that in the long term some of our worries may prove to have been unjustified, a shortage of unbiased information makes rational judgements very difficult. Clearly, it is too much to expect the food industry to offer unbiased advice, though where there is competition between rival foods, health arguments are likely to be given a better airing than in a monopoly situation. Thus butter and margarine manufacturers knock each other's products, but nobody attacks sugar. The bakers contrive to have it both ways, with Allied Bakeries promoting Allinson's wholemeal bread on one hand and squashy Sunblest on the other, while Rank Hovis McDougall extol the virtues of Windmill high fibre white bread – which first has all the natural fibre removed, and then has vegetable fibre from peas incorporated to remedy the deficiency.

Regrettably, the structure of the government's own food policy mechanism does little to inspire confidence in its impartiality. In a recent paper on food additive controls in the United Kingdom, France,

West Germany and the United States, Abraham and Millstone remark that, 'In each of the countries under review there are government departments responsible for additive regulations and for the sponsorship of the food industry. The UK is the only country in which the performance of both these functions is delegate to the same ministry. Such an arrangement would not be considered acceptable in the USA, France or FR Germany'.[9]

The overall picture is not quite so simple. The major responsibility for food policy in Britain, since the demise of the wartime Ministry of Food, has been the Ministry of Agriculture, Fisheries and Food (MAFF), which enjoys relations with the food industry almost as cosy as those with farmers commented on in the previous chapter. A minority interest still resides with the Department of Health and Social Security (DHSS). Usually the two ministries get their act together very well, as occurred when an embarrassed government wished to stall publication of the second report of the Advisory Committee on Irradiated and Novel Foods until after the 1987 election. In response to journalists who wanted to know when the report would be published, representatives of MAFF insisted that DHSS would be making a statement soon, while DHSS spokesmen were equally adamant that the announcement would come from MAFF.[10] Occasionally this concord is lost, as in the salmonella in eggs fiasco, when representatives of the two ministries were regularly to be seen on television contradicting each other in ever-more sweeping terms.

In forming their judgements on matters of food safety and policy, the men from the ministry are assisted by a number of expert committees: the Food Advisory Committee (FAC), the Committee on Toxicity (COT), the Advisory Committee on Pesticides (ASP), the Committee on Medical Aspects of Food Policy (COMA), and so on. Because a lot of food research is industry-funded, it is inevitable that many of the experts on these committees are sympathetic to the interests of the food industry. The difficulties of just policy-making are compounded by the vested interests of many MPs. In his fascinating study *The Politics of Food* Geoffrey Cannon documents the links between MPs, government advisors and the food industry, and lists 250 members of parliament who have some links with 'unhealthy' food businesses. He also describes the secrecy which makes such links difficult to discover, and which make objective outside judgements practically impossible, quoting with obvious approval from Hannah Arendt's *Crisis of the Republic*:

> Secrecy – what diplomatically is called discretion, as well as the *arcana imperii*, the mysteries of government – and deception, the deliberate

> falsehood and the outright lie used as legitimate means to achieve political ends, have been with us since the beginning of recorded history. Truthfulness has never been counted among the political virtues, and lies have always been regarded as justifiable tools in political dealings.[11]

Such cynicism may be unduly pessimistic: given a reasonable level of public interest, governments may begin to take healthy eating more seriously. During the Second World War, rationing provided a frugal but adequate diet for all as a matter of public policy. Soldiers have long been known to march on their stomachs, and the War Cabinet realized that the productive effort at home could only be effective if the workers were well fed. In more relaxed times, ration books are not an option, and the power of government propaganda is greatly diminished. The nutrition advertising budget of the Health Education Council was £142,000 in 1985; this compares with £483 million for the food industry as a whole, of which Mars Ltd spent around £35 million.[12] Food had effectively ceased to be a political issue until the worries of the medical profession over rising premature death rates and their possible dietary causes eventually began to cause public alarm. Because of the general complacency of the post-ration-book era, government policies on food quality had confined themselves to the essentially negative function of prohibiting the worst forms of adulteration and fraudulent labelling. There are now some signs of a change in attitude, with the announcement of a new Food Safety Directorate to take control of all aspects of food safety within the Ministry of Agriculture. But this move, and the government's plans to spend an extra £1 million on salmonella research, have not convinced sceptics, who point to concurrent plans to close the Bristol laboratory of the Institute of Food Research, which deals with meat and poultry hygiene.

The British policy of agricultural self-sufficiency, adopted even more enthusiastically after we joined the European Community, was motivated by worries over the balance of payments and the interests of farmers and land owners, and had nothing at all to do with quality. Indeed, the Common Agricultural Policy inevitably encourages quantity rather than quality. In a free market, when there is a glut the farmer with the worst quality produce should find it difficult to sell at a profit, whereas with a guaranteed market for all he can sell there is little incentive to maintain quality above the bare minimum levels set by bureaucratic decree. This is particularly true where competition at the retail level is stifled by monopolistic production or marketing arrangements. The insipid uniformity of the Milk Marketing Board's Dairy Crest cheeses is a case in point. Conversely, when there is no intervention price, or only a very low one, the pursuit of quality becomes worth while. The extreme example of this can be found among wine growers, where those with the best product achieve very much higher prices.

Curiously enough, it is the supermarket chains, for all their associations with packaged and processed foods, that have been most involved in such improvements in food standards and labelling as have occurred in the past few years. The major food retailers, turning over a billion pounds or more, are obvious targets for consumer pressure groups. They are also conscious of their national public image, and sensitive to the effects of the adverse publicity that prosecution for unfair trading could bring. In a survey of fish shops in Preston, environmental health officers discovered that only two out of forty samples of smoked fish declared the presence of added colouring material. This is hardly national news, but if supermarket produce were found to be similarly mislabelled there would be considerable media coverage, and rapid action by the stores to reassure the public that the problem would not recur.

Prompted by consumer organizations and media concern, the big supermarket chains have put pressure on manufacturers to cut down on food additives and to label products adequately, often insisting on much higher standards than the law demands. As well as selling free range eggs, Marks and Spencer have introduced free range poultry products in a number of their stores, and are encouraging their suppliers of pig meat to evaluate more humane systems. At present, supplies of naturally reared meat are very limited, and because of the small scale of most of the producers it is difficult for supermarkets to maintain the level of inspection required to be sure that what they are selling is exactly as advertised. Organic vegetables have been more successful, and Safeway now offer a range of twenty to thirty types of organically grown produce. Organic produce now accounts for 5 per cent of retail fruit and vegetable sales, and would be more if supplies were available; Safeway complain that up to 60 per cent of fresh organically grown produce has to be imported, mainly because of difficulties convincing farmers in this country of the growing demand.

There has been less progress regarding the problem of residues in food. While the big retailers do extensive testing for pesticide residues in fruit and vegetables, they have so far been unwilling to provide the public with any information on residue levels. No doubt this is partly because they vary so much, and it would be necessary to test virtually every item for sale before maximum residual quantities could be guaranteed. Similarly, while a firm like Marks and Spencer can lay down detailed feed specifications for poultry reared in large broiler farms, it is impossible to guarantee the absence of hormone or antibiotic residues in beef bought at market without testing every single carcass.

Manufacturers and retailers find advantage in the processing of food to give it a longer shelf life. Whole-grain flour turns rancid more quickly than white flour; fresh fruit rots but processed sugar is a commodity that keeps indefinitely and can itself be used as a preservative. But with the growing consumer interest in fresh food, the supermarket chains have been quick to adapt, and have exploited their advanced computerized distribution networks and rapid turnover to give them a competitive advantage in quality as well as in price over the little corner shop where most of the groceries still come in packets or tins.

Supermarket shopping offers consumers a freedom of choice which is quite extensive in the more 'up market' stores. Few people would wish to have a healthier diet imposed on us by the banning of all supposedly unhealthy foods, or by rationing how much of particular items we are allowed to eat. On the other hand, we may need some protection from the power of the food industry's chemists to deceive our palates, and of its public relations to confuse our judgement. Consumer pressure has persuaded the present government to somewhat improve the nutritional labelling of food, though the format in which the information is presented is so variable as to make comparisons between products very difficult. At the fashionable end of the market, fresh produce has begun to be distinguished more by variety, and the day may come when customers at Marks and Spencer can choose Charolais, Hereford or Angus beef as well as Desirée, Maris Piper or King Edward potatoes. In this consumer bracket, producers have nothing to fear from labelling, and the Soil Association wisely defends its standards against opportunists who would debase the 'organic' name-tab, with all the fervour that famous châteaux in France defend their *appellations controlées*. But labelling cannot be the whole answer. For social security shoppers, a potato is just a spud and beef mince is all too often a concoction of nameless scrags. In this milieu, suppliers resist labelling or make it deliberately confusing, the brand name is everything, and junk food reigns supreme. The poorer consumer will not be protected by nutritional information in small print, and a policy of *caveat emptor*. Minimum food standards are the only answer, not only regarding permitted additives, but also for the nutritional content of both fresh and processed foods. It may be too late to save any meaning at all for such abused words as 'fresh', and it is probably irrelevant whether Cadbury's Dairy Milk qualifies as 'chocolate' under EC nomenclature; but it is surely important that fresh meat should contain less than 75 per cent water, and that all processed foods should have as near as possible the same nutritional value as their fresh equivalents.

The general public do not just suffer from current agricultural policy as consumers paying too much for inferior produce, but also as users of the countryside suffering the loss of amenity value which modern farming methods can cause. A majority of people make at least occasional visits to the country for pleasure, while an increasing and influential minority choose to commute to work from country homes. To both these classes, the fresh air, quiet and natural beauty of the countryside are attractions of significant value; and it would be pointless, and a denial of the environment conscious spirit of the age, to argue that only the farming community has any real right to the benefits of country life. Since their forebears of three hundred years ago planned, executed and got away with, the great enclosures of common land which shaped the rural landscape of today, farmers have enjoyed huge political influence. Before Disraeli, farmer control of the British Conservative party was absolute, and it is only in the past few years that it has seriously been questioned whether the agriculture industry deserves its unique seat at the Cabinet table. In 1987 the National Farmers' Union set its strength against the authority of the Minister of Agriculture Michael Jopling, and demanded his replacement. Jopling's head rolled, but the NFU emerged as the losers. His successor, John MacGregor, was far less sympathetic to the farmers; and the Conservative backbench Agriculture Committee gave the NFU leader Simon Gourlay a dressing down, telling him that the days of special treatment for farmers were over.[13]

One of the main arguments which led to this *dénouement* was over alternative land use in the rural economy, and the Department of the Environment's bid to extend its influence in countryside planning. Farmers have traditionally occupied a supremely privileged position regarding planning control, with virtual exemption from local government interference, and complete freedom from rating. The NFU regards farmers as 'the natural custodians of the countryside', but their 'Agenda for Agriculture' makes it abundantly clear that if they are expected to farm in an environmentally friendly way, they had jolly well better be paid for it. Unlike other industries, whose environmentally unacceptable practices can be legislated against, and whose projects require planning permission if they materially alter the landscape, farmers demand the right to build, plough, plant, cut down or burn without any qualification. They are, in fact, willing to be bought off with a kind of danegeld made available under the 1981 Wildlife and Countryside Act; so that a

Dorset farmer can get £20,000 per year for the next sixty-five years for not uprooting woodland to grow wheat, or four Norfolk farmers can pocket £2 million over the next twenty years for not draining part of the Halvergate marshes.[14] In respect of pollution of watercourses, farmers are treated more or less on a par with other industries, but they again have a privileged position when it comes to the disposal of toxic chemicals. The present codes of practice are absurdly lax, simply allowing more or less any poisonous substances to be buried in shallow pits on farmland.[15]

As the public becomes less convinced by the arguments for the special status of agriculture, it is to be expected that farming will become more exposed to planning controls. The most offensive aspect of agriculture from the point of view of loss of amenity is probably the operation of intensive livestock units. These tend to be visually intrusive, noisy, and above all, smelly. Under class VI of the 1977 General Development Order, intensive livestock units were eligible for exemption from planning permission as 'operations requisite for the use of that land for the purpose of agriculture'. In practice, if the available land was capable of growing less than 35 per cent of the fodder requirements of the housed livestock, the Ministry of Agriculture required planning permission before granting aid to intensive livestock units, but the owners of larger holdings were generally allowed to erect livestock houses anywhere right up to their boundaries. In the pig farming region of Humberside, the district councils produced 'guidelines' proposing that intensive livestock units (and slurry spreading operations) should be kept at least 100 metres from settlements, and 800 metres from urban areas.[16] These proposals followed the pattern of restrictions in many other countries; Greece, for example, had similar regulations already in force by 1977.[17] In a review of the General Development Order starting in 1984, the exemption of livestock buildings within 400 metres of residential properties, schools or hospitals was abolished, and slurry was added to the list of substances requiring consultation with the water authority before development can proceed. The Yorkshire Water Authority had already produced guidelines proposing that there should be no slurry application within an area of 3 square kilometres around a borehole used for public water, or within a screening zone of 1½ square kilometres upstream of a surface water intake.

The gradual subjection of farmers to the same sort of planning controls as the rest of us is to be welcomed, but even if the planners allow factory farm developments, it is certainly the case at present that farmers are not bearing the full costs of agricultural intensification in terms of pollution control and general loss of amenity. This is particularly unfair in view of their total exemption from rating, which was challenged by a number of local authorities who sought to rate factory farm buildings, but was subsequently reinstated by Parliament under pressure from the farming lobby. Peter Roberts points out that a light industrial unit the size of an average broiler unit might be expected to pay £12,000 per year in local rates; whereas the multinationals who own egg batteries or intensive pig units are simply cashing in on the public's willingness to supply policing, fire services and roads for free.[18]

Open air feedlots are just as offensive as battery houses, both to the eye and to the nose. The first large feedlot in Britain was established in 1987 by Frans Buitelaar (Farms) Ltd, for fattening up worn out dairy cows for slaughter. Huge earthworks and attendant access roads were constructed in the Lincolnshire Wolds without any application for the necessary planning permission. Buitelaar made a late application for permission to cover the work already done, but hastily withdrew it when the District Council's planning sub-committee voted to turn it down in September 1988. This meant that enforcement notices had to be issued, followed by an appeal and a public enquiry, which eventually started in October 1989. Extensive damage has meanwhile been done to the country roads near the site by heavy vehicles transporting cattle and fodder, and in 1988 Buitelaar were successfully prosecuted by Anglian Water Authority for pollution offences. The feedlots (still in operation at the time of writing) accommodate several thousand cattle, which stand or lie in 'a sea of black mud', and are fed on wet mash compounded from vegetable wastes and poultry manure. Although the cows may not exactly be happy, there is no evidence that Buitelaar neglects them: in fact, their managing director is on the government's Farm Animal Welfare Council.[19]

Unlike their land-based colleagues, fish farmers do not claim a hereditary right to preferential treatment under the planning laws, but the fish farming industry has grown up so fast that many of its activities have been completely beyond the scope of contemporary legislation. A chaotic situation resulted, in which local authorities and river boards were uncertain of the extent of their regulatory powers, and it is only recently that the problem of regulating the development of offshore fish farming has seriously begun to be addressed.

In Britain, the offshore sea-bed is regarded as a hereditary possession of the Sovereign, and under the Crown Estate Act of 1961 its management is in the hands of the Crown Estate Commissioners, who have a duty 'to maintain and enhance its value and the return obtained from it with due regard to the requirements of good management'. Until 1986 the Crown Estate Commissioners were the sole arbiters of sea-bed lease applications unless there were objections from the Department of Transport on navigational grounds. While consultations with other interested parties were sometimes made, the Commissioners actively encouraged lease applications, to maximize the income received from their tenants, and about 350 leases were granted under these arrangements before 1986. The offshore activities of fish farmers are immune from local authority planning control, and following critical comments by a number of voluntary and official bodies, the Scottish Development Department drew up a consultative procedure for the Crown Estate Commissioners, which has been in use since October 1986. In his Foreword to the Commissioners' 1987 report, the Earl of Mansfield welcomed the new procedures as enjoying 'widespread support'. Under the new system a number of statutory and voluntary bodies are given twenty-eight days to comment on individual lease applications, after which the Crown Estate Commissioners make the final decision. For all Lord Mansfield's confidence, the response from those who are neither Commissioners nor directors of fish farms has been less than ecstatic. Local authorities object to their subordinate role in the consultation process; the Nature Conservancy Council take exception to the continued presumption in favour of development and the absence of any coherent environmental planning strategy, and the Scottish Wildlife and Countryside Link point out that the planning control exercised by the Crown Estate Commissioners 'almost by default' is fundamentally at odds with the normal democratic requirements of impartiality and accountability.

Since the first fish cages appeared offshore, there have been moves by a number of other statutory bodies to bring the new industry under their regulatory control. After taking legal advice, the Highland River Purification Board decided that fish cages required consent to discharge waste under the 1974 Control of Pollution Act, and since they adopted this policy in 1987 the other river purification authorities have followed suit. A major problem at present is their lack of resources to carry out an effective monitoring policy. Under the Diseases of Fish Act 1983, all fish farms must now be registered with the Department of Agriculture and Fisheries for Scotland, who are also responsible for issuing licences to shoot seals and birds. In 1988 the Environmental Assessment (Salmon Farming in Marine Waters) Regulations gave the Nature Conservancy Council an improved opportunity to oppose certain fish farming developments, but left the final say to the Crown Estate Commissioners.

Fish farm installations on land do not enjoy the same immunity from planning control, though there is some confusion in the interpretation of the General Development Order, regarding whether or not fish farming is a form of agriculture. Most Scottish local authorities have now adopted the position that all landward developments relating to fish farms must seek planning permission, and recently this policy has been used with effect to compensate for the planning authorities' lack of jurisdiction over offshore sites. After a public inquiry in September 1989 the Deputy Chief Reporter at the Scottish Office dismissed appeals by a fish farm company against refusal of planning permission for two shore-based developments on Loch Long. The Reporter pointed out that the granting of a sea-bed lease was no reason for a presumption in favour of planning permission for shore bases; nor could he see why the landscape and other environmental aspects of the cages themselves should not be taken account of in the determination of planning applications.[20] This judgement further undermines the autocratic powers of the Crown Estate Commissioners, but with some 500 sea-bed leases already in force, it comes rather late in the day.

It is normal business practice to pursue profits within the constraints set by the law; it is even common enough business practice to see how far the law can be stretched before it becomes a constraint, so if farmers are businessmen rather than general benefactors, we should not be surprised at conflicts between their interests and those of the public at large. We should, rather, be suspicious when everything in the garden seems too attractive: it is arguable that most of the problems discussed in the two preceding chapters have been made worse by uncritical public acquiescence. Corrective action, whether related to the Common Agricultural Policy or matters of local planning, has often been both too little and too late. The economic and public health questions raised above can in general be reduced to decisions between the selfish interests of the farmers or food manufacturers on one hand and the public on the other. But this reduction to a conflict situation is not always practical or worth while. If the conflict is between our present demands as consumers and the interests of Third World farmers or of future generations, then our selfish concerns have little to fear from such distant competition. Likewise, however much our society abuses farm animals it is highly unlikely they will ever come out of their cages to get us. In these cases, sentiments of compassion may inspire action, while feelings of self-interest will not. The subjects of the three final chapters will be the evolution of a wider concern for animals, the resulting reforms and their international implications, and the prospects for further progress.

Part VI

PROSPECTS FOR REFORM

15

Battle Joined

Sympathy for other animals is not a new phenomenon. The Roman historian Plutarch and the neo-platonist philosopher Porphyry both wrote at length in favour of vegetarianism; or if we must kill an animal, Plutarch urged, 'let us do it with sorrow and pity and not abusing or tormenting it as many nowadays are wont to do'.[1] In all ages there have been sensitive thinkers who abhor cruelty to animals, as well as sadists who revel in the gratuitous infliction of pain; but how much we can know about the general attitudes of previous generations is open to doubt. It is as certain that French peasants did not all start throwing cats into the fire because they had read Descartes, as it is that their English counterparts ignored Blake's advice not to kill flies and moths. On the whole, domestic animals have been treated like other possessions; looked after by the thrifty, neglected by the careless – and disposed of when they were no longer wanted. Some have been loved, like the lamb in Nathan's parable, and many have been misused cruelly. Most have been cared for by masters who when the time came, have also been their executioners. James Serpell has given a fascinating account of the ambivalent attitudes towards animals in a variety of cultures.[2] Exploitation of other species demands understanding, and with understanding comes sympathy. The inevitable conflict of feelings finds expression in the way that man can 'switch off' the affection that has grown up with years of close association, and callously subject his beasts to horrible tortures in their final hours.

While personal feelings towards other species are characterized by variety and ambivalence, the attitudes of most societies as reflected in their laws can only be described as indifferent. In the ancient civilizations of India and Egypt there were severe penalties for killing various types of animal; but, as in many more primitive cultures, these restrictions were founded on religious beliefs which are now generally inaccessible to us. From Roman times until the nineteenth century, legislation throughout Europe has dealt with animals only as property, and has consistently ignored the question of cruelty. Bearing in mind the mass of detailed, trivial and interfering laws which have been passed during the period in question, the degree of tolerance regarding conduct towards other animals is quite outstanding. Of course, even minor details of human conduct can be perceived as threats to the social fabric, whereas the treatment of dumb animals is unlikely to be seen in this light. By the sixteenth century, the struggle for survival and the religious orthodoxy of the Middle Ages were being replaced by a new self-confidence and a desire to question the established order; and the Renaissance brought the question of man's relationship with other species back into the realm of moral and theological debate. Whether this had any effect on the way animals were treated by the populace at large is doubtful: although reports of cruelty become much more common from Tudor times, this is almost certainly due to renewed interest in the subject among intellectuals. The arguments were not all on the animals' side; for while such Humanists as Thomas More and Montaigne railed against cruelty, any breaking down of the species barrier was seen by others as a threat to the religious doctrine of man's dominion over the rest of nature. By the time of Descartes, the arguments for man's uniqueness were becoming strained – though the great philosopher managed to save his faith by insisting that while animals were mere machines, man's possession of an immortal soul left him in a completely different category. This was a convenient belief for those early animal experimenters who wanted to both have their cake and eat it: they could inflict cruel tortures while denying that they were cruel, because a mechanical creature with no soul could not possibly feel pain.

The major philosophers of the European Enlightenment were unimpressed by this line of special pleading; and one by one Voltaire, Rousseau, Hume and Bentham reiterated the argument that what we share with other animals, in nature and in society, obliges us to treat them with compassion. In literature, too, many of the major figures took up the humane cause: Milton, Pope, Blake, Wordsworth, Coleridge, Byron and Shelley all expressed concern for the suffering of other creatures at the hands of men. By the end of the eighteenth century, civilized opinion was against the wanton infliction of pain on brute beasts, but the butchers, drovers, bear baiters and huntsmen still carried on their activities pretty much as they had always done.

The first British legislation to protect animals was enacted in 1822. The Act to Prevent the Cruel and Improper Treatment of Animals was carefully drafted in terms restricting its scope to the cruel treatment of animals belonging to other people, thus avoiding the question of whether men could do what they like with their own, and making it clear that its target was the working class of drovers and carters rather than the influential aristocracy. It was introduced by Richard Martin, the witty and extrovert Member of Parliament for County Galway, who had in his youth fought a duel over the shooting of a friend's dog, and whose generosity to beast and man led George IV to nickname him 'Humanity Dick'. Martin himself brought the first prosecution under the act, producing as evidence in court the donkey whose ill-treatment was the subject of the charge; and he continued to publicize the new law with a number of similar cases, sometimes paying the miscreants' fines himself rather than cause them undue hardship.[3]

It was to be a long struggle for the reformers, with many subsequent attempts at legislation being thrown out or watered down beyond recognition. In 1824 the Society for the Prevention of Cruelty to Animals was formed, and under Queen Victoria's patronage it gained the 'Royal' prefix in 1840. By the end of the century, statutes were in force banning all animal fights and giving a measure of protection to wild birds, laboratory animals and livestock in transit. The RSPCA employed inspectors to ensure that the new legislation was enforced, and issued numerous tracts and pamphlets in an effort to educate the working class. The successes which were achieved owed a great deal to the reformers' backing by an impressive array of influential and titled supporters, but this also meant that the cruel habits of the influential and titled were let off relatively lightly compared with those of the lower orders. To this day the RSPCA have failed to achieve the abolition of hunting, which many of their rank and file would wish to see; while animal experiments by learned professors and cruel but profitable farming practices have only been curbed to a limited extent.

The achievements of the Victorian era were consolidated in Sir George Greenwood's 1911 Protection of Animals Act, which still covers many aspects of the treatment of farm animals. Under this Act, anyone who causes 'unnecessary suffering' to an animal is guilty of an offence of cruelty. In 1933, the pre-stunning of cattle at slaughter became a legal requirement, and further acts extended this protection to pigs and sheep, though exempting the ritual slaughter practised by Jews and Muslims. A number of more recent laws such as the Agriculture (Miscellaneous Provisions) Act 1968 and subsequent Orders made under this Act, lay down in more precise detail the conditions under which various operations such as branding, tail-docking and castration can be performed, and specify regulations for the transport of animals by road, rail, sea and air.

There is no doubt that the animal welfare laws, with the RSPCA playing a major role in their enforcement, have abolished a number of painful and injurious practices which used to be widespread on farms and when the animals were taken to market and slaughter, and have reduced the extent to which farm animals are subjected to deliberate cruelty. But have they made the lives of the majority of farm animals healthier and happier? Unfortunately, though neglect and wanton maltreatment have almost certainly become less common, there is an opposing tendency for livestock in modern farming to be ever-more deprived of a remotely natural life style. As farm livestock, the majority of hens and pigs are kept in conditions which would not be tolerated for a moment if they were kept in a zoo or as domestic pets. In 1964, Ruth Harrison brought the fate of animals in factory farms into the public view with the publication of her book *Animal Machines*. The government were prompted to set up a committee of investigation under the chairmanship of Professor Brambell; this recommended the prohibition of close tethering of veal calves and sows, as well as the debeaking of hens and the docking of pigs' tails, but no action was taken to make any of these practices illegal.

The RSPCA were shy of criticizing intensive farming methods. The sixties and seventies were a difficult period for the Society, as radical and conservative groups struggled for control of policy in a number of bitter confrontations which centred on the main issue of hunting, but also involved allegations and counter-claims over the Society's finances. A number of factory farmers were also members of the Society, and obstructed attempts to formulate an agreed policy on the subject. Although the RSPCA has now got its act together, by far the most effective campaigning for improved conditions for farm animals has been carried on by much smaller pressure groups which specialize in this field; of which Compassion in World Farming is the most significant. Compassion in World Farming was founded in 1967 by an ex-farmer, Peter Roberts, whose tireless efforts in the face of many disappointments deserve great admiration. The cause of battery hens has been championed with equal determination by Violet Spalding and Clare Druce, who set up the catchily-named Chickens' Lib in 1971.

Following the report of the Brambell Committee, which recommended among other things that 'an animal should at least have sufficient freedom of movement to be able without difficulty to turn round, groom itself, get up, lie down and stretch its limbs', the government set up a Farm Animal Welfare Advisory Committee. This Committee helped the Ministry of Agriculture to draft Welfare Codes 'intended to encourage stock-keepers to adopt the highest standards of husbandry'. The welfare codes were to be advisory rather than mandatory; in contrast to animal health regulations, which are enforced by fines sometimes running into tens of thousands of pounds. The first codes were issued in 1971, and revised versions were produced from 1983 onwards by the reconstituted Farm Animal Welfare Council (whose members, like those of the earlier committee, were a mixture of academics, representatives of animal welfare interests and directors of major farming concerns). The most recent codes acknowledge that welfare involves more than the absence of injury or disease, but relates also to important behavioural needs, such as:

freedom from thirst, hunger or malnutrition;
appropriate comfort and shelter;
prevention, or rapid diagnosis and treatment, of injury and disease;
freedom to display most normal patterns of behaviour;
freedom from fear.

According to the 1968 Agriculture (Miscellaneous Provisions) Act, failure to observe Welfare Code provisions is not itself an offence, but if a farmer is prosecuted under section 1 of the Act for causing 'unnecessary pain or unnecessary distress', such failure 'may ... be relied upon by the prosecution as tending to establish the guilt of the accused'.

The welfare codes were widely criticized when they were first introduced, for being too vague, for not going far enough, and for being unenforceable. Despite the obvious intentions of their authors, there was certainly little sign at first of the codes having any impact on established farming practices. In 1976 the National Society Against Factory Farming prosecuted a large firm of veal producers for failing to abide by the 1971 Code, and although they did not secure a conviction, the producer soon afterwards changed over to the more humane straw yard system, which has proved equally successful in practice. A similar case brought by Compassion in World Farming in 1984 to test the new welfare code also failed, and the defendants were awarded costs of £12,000. Tenacious lobbying following this case finally convinced the government that voluntary codes of practice were not enough and that enforcible welfare regulations were needed.

When the first mandatory welfare regulations were announced, they proved to be somewhat disappointing. Compulsory display of the welfare codes, prohibition of sharp projections which would cause injury to stock, and the provision of clean drinking water for all stock struck many welfarists as fairly minimal steps in the right direction. There was still no mention of an animal needing adequate space to turn round, no requirement to provide a dry lying area, no prohibition on solitary confinement, no stipulation regarding the need for exercise, and no mention of fire precautions. The new regulations were not welcomed by the National Farmers' Union, who saw them as the thin end of the wedge of further legislation to control production methods. Indeed, in the same year, the government did announce its intention to ban veal crates from January 1990, just twenty-five years after Brambell first recommended they should be abolished.

Veal is a minority taste in Great Britain, so the veal industry was a fairly soft target for welfare legislation. The multi-million pound bacon and egg farmers have a much more powerful lobby to beat, and not just in Westminster, as will become clear in the following chapter. As economic arguments in favour of less intensive methods of pig farming gather force, and as the welfare codes continue to condemn dry-sow stalls in increasingly strong terms, there are signs that pig breeders have seen the writing on the wall, and are beginning to change over to more humane systems. But the battery hen operators are likely to prove a much harder egg to crack, since their costs are inevitably much lower than those of free range producers. The egg industry already spends a lot on advertising, including the provision of glossy information leaflets for schools, and it is determined to suppress the opposition if at all possible. Whether or not the battery farmers are directly to blame, Compassion in World Farming (CIWF) has twice suffered major set-backs in its attempts to publicize the plight of caged hens. In 1987 CIWF were offered the chance of a full page advertisement in *Reader's Digest*, the magazine giving one of its charity pages and the advertising agency Burkitt Weinrich Bryant (BWB) doing the design work, and all completely free. But in April 1988, just a fortnight before the advertisement was due to appear, the *Reader's Digest* refused to accept it, in case the picture of a chicken confined in a small space might cause offence. Just who might be offended became clear when CIWF placed the same advertisement in the magazine *Bella* – this time at their own expense. The Code of Advertising Practice Committee received complaints from the British Egg Industry Council, the UK Egg Producers' Association and the British Egg Information Service, protesting that not all of the 45 million hens in the country have festering sores or brittle bones, and that the hen was shown on its own in what appeared to be a single cage of inadequate size. Nor did Compassion in World Farming do much better with their 60 second *Welcome to the Battery* cinema advertisement. In this case it was members of the National Farmers' Union who persuaded the Cinema Exhibitors' Association to ban the advert, although it had been approved by the Cinema Advertising Authority and had been passed by the British Board of Film Censors. Following pressure from a number of town councils and members of parliament the ban was eventually relaxed, but inevitably the advertising campaign had lost some of its impetus.[4]

That freedom of information is seldom encouraged by commerce is nicely illustrated by the fur trade's successful scuppering of Alan Clark's Fur Labelling Order. This was introduced in Parliament with a fanfare of publicity, and required furs made from certain species to bear a label stating that the particular type of fur was normally obtained by use of leg-hold traps. The fur trade was incensed, and as well as threatening the government with court action, they organized an intensive lobbying campaign in Britain, Canada and the United States. This included letters to the Queen and Prime Minister from the Bishop of the Anglican Diocese of the Arctic, and from Canadian schoolchildren to British Members of Parliament. More significantly, Canadian diplomats, who were understandably reluctant to see their fur trade with the European Community go the same way as their wheat and dairy exports had gone, hinted that Britain might lose a valuable contract for nuclear submarines if the Order were not dropped. And so it was, like a hot brick.

Despite industry pressure, however, the animal welfare issue is still getting quite prominent media coverage. Television documentaries on factory farming have increased public awareness, as did the *Animal Squad* series about the work of the RSPCA. Newspapers, too, have given considerable space to farm animal welfare, and in November 1989 *The Sunday Times Magazine* featured a battery hen on its front cover – very similar to the one which the *Reader's Digest* had rejected just eighteen months earlier.[5] But whether there is really a swing in public opinion which will lead to further legislation is a moot point. Public opinion has never actually been keen on factory farms: in 1968 a Gallup poll found that 87 per cent of people thought farm animals should be able to turn round freely in their pens, and 91 per cent agreed that all birds should have room to spread their wings; while twenty years later a National Opinion Poll found that 82 per cent of the public agreed with both of the stronger propositions that 'it should be illegal to keep pregnant pigs confined in cages in which they are unable to turn around', and 'hens should not be kept in battery cages in which they are unable to spread their wings fully'.[6,7] These reactions of people who are actually asked what they think, and the substantial number of petitions and letters to Members of Parliament on animal-related issues, give the impression of a high degree of public concern. When pushing a supermarket trolley, however, the average consumer still shops very much by price – though a small and growing number of people are now prepared to pay extra for free range.

Such inconsistencies seem to be part of human nature: however willing the spirit, the weakness of the flesh prevails. It is easier not to think too deeply; indeed, there is a great temptation to arrange society so that we do not need to face up to our own inconsistency. Thomas More's Utopians did not eschew meat, but their butchers were an untouchable class who plied their trade in secret. The amiable Gilbert White planted a row of lime trees by his house at Selborne to conceal from view the butcher's yard opposite, and Dr Johnson remarked with embarrassment that 'he was afraid there were slaughterhouses in more streets of London than one supposes'.[8] So, too, do we both pity and arrange for the discreet exploitation of farm animals. Nevertheless, there are signs of change: the number of vegetarians, though still just 3 per cent of the population, is growing steadily, as is the demand for free range farm produce.

Farmers themselves are frequent critics of the methods they have to adopt, though they are defensive about their 'independence' – something of a myth in this age of subsidies – and resent outsiders telling them to put their house in order. Broiler farmers and veal farmers criticize battery egg producers, and battery farmers criticize those who keep their sows in stalls. When a cross-section of farmers were polled by Gallup in 1968, they were almost as generally opposed to close confinement systems as were the public at large.[9]

The intensive farming lobby is much wider than just the farmers themselves, extending through feed suppliers and the manufacturers of cages, equipment and agricultural chemicals to the large number of research workers whose jobs are tied to agro-technology. Science in our society has such an impressive record of delivering the goods, that we tend to assume that the high-tech solution is inevitably worth pursuing. Sometimes, though, the technological path leads to unacceptable social costs, or irresolvable ethical dilemmas, as many would argue has happened with intensive farming. Sometimes, the 'solution' is totally irrelevant, as there was no problem to start with: thus boosting milk yield with the synthetic hormone BST looks pretty silly to everyone except the Monsanto Chemical Company in the situation where milk production is being kept down to an acceptable level by the use of strict quotas. As a higher proportion of research is funded by industry, it is natural that scientists who want to push back the frontiers of knowledge in exciting ways will become ever-more anxious to prove that their discoveries can make money – even if it is by satisfying a hitherto unthought of need. Most academics are conscious that such sentiments should not be voiced in public, but occasionally they allow their worries to get the better of their discretion. At a recent poultry conference, for example, one professor was in favour of a public relations campaign to promote intensive farming methods, because otherwise 'the political future of all of us in intensive animal production is going to be very dicey'.[10]

Sadly, the group which has got itself into the worst cleft stick regarding animal welfare is the veterinary profession. All vets take an oath on graduating that they will constantly endeavour to ensure the welfare of animals committed to their care, and there is no doubt that the profession contains a great many dedicated practitioners who do exactly this. But some of the high-flyers get to the top by pursuing research interests which many critics regard as unethical; and the demands of legislation mean that others spend their lives in the morally desensitizing environment of the animal testing laboratory, the slaughterhouse or the factory farm. Since in any of these situations the animals involved are invariably under the nominal care of a veterinary surgeon, there is understandable reluctance on the part of other members of the profession to be too outspoken in their public condemnation. When cruelty cases come to court it has been suggested that the vets' fear of public disagreement, in what is a small and tightly knit profession, may impair their value as witnesses.[11] Under existing arrangements for the veterinary service, there is an inevitable danger that the farmers who pay the piper also expect to influence the welfare tune he plays. Professor John Webster recently wrote to the *Veterinary Record* on this subject, calling on practitioners to ask themselves if they were achieving the right balance between animal welfare and the desire for an adequate income.[12]

Finally, opposition to reforms can simply be the result of innate conservatism and a reluctance to change anything at all. The small minority of violent campaigners for animal rights provoke this sort of reaction; and Richard Ryder records how the radical reformers of the RSPCA were repeatedly accused, quite unjustly, of having communist sympathies.[13] The farm animal welfare campaign has largely escaped this sort of opprobium, thanks partly to the responsible policies of groups such as Compassion in World Farming, and partly because fur shops and laboratories offer the extreme militants more glamorous and 'softer' targets than chicken sheds and piggeries. The growing respectability of the cause was emphasized in 1984 when the Governor of the Bank of England urged newcomers to stock farming to be less intensive.[14] In a recent South African novel, the remark occurred that once the banks had liberalized, you could be sure the Church would not be far behind. The recent announcement by Dr Runcie of a Church of England Commission on Rural Affairs, with intensive farming as part of its brief, must surely lend credence to this memorable maxim.[15]

16

The International Scene

Soon after Richard Martin's 1822 Act, other European countries began to follow Britain's lead and give animals a measure of protection under the law. By 1860 Germany, Switzerland, Norway, Sweden and Denmark had all passed welfare legislation, and several countries had founded animal protection societies along the lines of the RSPCA. In America, one or two states passed anti-cruelty laws in the 1830s, though it was 1866 before the American Society for the Prevention of Cruelty to Animals was founded by Henry Bergh, also taking the RSPCA as its model. In 1884 Bergh was asked to help a 'little animal' which was being tormented by a woman; but the animal turned out to be a human child. Undeterred, he managed to prosecute the woman successfully, and through this and subsequent similar cases the New York Society for the Prevention of Cruelty to Children was formed. This initiative was quickly reflected on the eastern side of the Atlantic, where the British RSPCA committee were instrumental in founding the National Society for the Prevention of Cruelty to Children.

During the reign of Queen Victoria, animal welfare societies were founded in numerous countries throughout the British Empire, and in the twentieth century Britons have continued the tradition by helping to set up similar organizations in Greece, North Africa and Italy. The countries of northern Europe have generally speaking kept pace with each other in the field of welfare legislation, with the United States of America lagging somewhat behind. The first federal humane slaughter laws were enacted as recently as 1958, and Americans on the whole remain relatively unaware of the problems of farm animal welfare.

Outside the industrialized Western world, animal welfare laws remain primitive or non-existent. Even in Europe, the Latin countries lag far behind those of the North, and their animals have only minimal protection. Spain had numerous humane societies at the end of the nineteenth century, and by 1930 the Ministry of the Interior presided over more than 4000 local animal and plant protection committees. A number of welfare laws had been introduced, and a humane education programme started among schoolchildren. With the support of General Franco, the Church opposed these developments as 'foreign ideas and protestant influence', and after the Civil War there was no further progress until the revival of Spanish democracy in the 1970s.[1] In Italy and Greece, there has been virtually no legal protection for animals until the implementation of recent EC directives.

In many countries throughout the world, animal welfare has yet to be recognized as a problem, far less legislated for. China, for example, has a law to protect wildlife, but none for house and farm animals. As in other Asian countries, methods of slaughter are primitive in the extreme: a sack is put over the head of the animal, which is beaten with a stick until it collapses, after which the throat is cut. The majority of the population simply do not regard animal suffering as important.[2] In 1950, the World Federation for the Protection of Animals was formed with the aim of promoting internationally the cause of animal welfare; and in 1959 the RSPCA and the Massachusetts SPCA set up the International Society for the Protection of Animals for similar purposes. These two bodies merged in 1980 to form the World Society for the Protection of Animals (WSPA). They have carried out valuable work rescuing wildlife threatened by the flooding of new dams, and promoting humane slaughter methods, and the WSPA is currently establishing links with China, Russia and other countries in the hope of persuading them to introduce animal protection laws along Western lines.

During the past ten years, several other European countries have gained a decisive lead over Britain in their farm animal legislation. Article 1 of the 1972 German Animal Protection Act is similar in tone to our own 1911 Act: 'This Act shall serve to protect life and well-being of the animal. Without reasonable cause nobody shall cause pain, suffering or injury to an animal'. But in the first paragraph of article 2, the German law went further, insisting that any animal must be given adequate food and care suitable for its species and accommodation which takes account of its natural behaviour; also that it must not be confined so that the needs of the species for movement and exercise are restricted to the extent of causing avoidable pain, suffering or injuries.[3] Under this legislation, a number of cases were brought in the German courts against battery egg producers, but with inconclusive results. While the courts might accept that caged hens were exposed to avoidable suffering, they were reluctant to convict the farmers of any offence. On the one hand, photographic studies have repeatedly shown that hens require far more space than they get in battery cages for activities such as preening or wing stretching, but on the other hand the defence could invariably produce expert witnesses who were confident that 'these inconveniences are not too much for the chicken' or that 'the chickens do not at all care whether they have feathers or not'.[4] Because of the readiness of some scientists and veterinarians to testify on behalf of the battery producers, the courts tended to take the view that if established experts are unable to agree on when and how the hens are suffering, the farmer can hardly be held responsible. Although the Frankfurt Court of Appeal ruled in 1979 that the confinement of hens in battery cages did in fact constitute cruelty, and the state of Hesse took steps to ban battery farms, the eventual result for the rest of the Federal German Republic was the amendment of the 1972 Act in 1987, to include detailed regulations which specifically permit battery cages subject to certain minimum standards.

Like the unsuccessful British prosecutions of veal farmers, the German experience illustrates the difficulty of achieving definitive court judgements on the basis of interpretation of rather general terms such as 'unnecessary suffering'. To avoid endless arguments, it may be better to frame more legislation in terms of precise situations, so that the court is concerned only with proof of the facts of the case, rather than with interpretation of the spirit of the law. Martin's 1822 Act specifically referred to beating, as well as other forms of cruelty, and the Spanish law still concentrates on forbidding particular abuses such as beating, overburdening and kicking. Similarly, much modern legislation tends to lay down regulations to cover such frequently recurring matters as slaughter techniques or the maximum intervals between feeding and watering animals in transit. Since it is obviously impossible to cover every conceivable situation, general principles are also required; but while in our country at least there would be a reasonable consensus on what is meant by cruelty, concepts such as deprivation, mental suffering or needs of the species are much more controversial in their exact meaning.

It is conspicuous that the countries which have recently made most progress towards the reform of intensive livestock farming have done so through the introduction of detailed regulations. Once these become law, there is less danger of the legislators' intentions being whittled away in local courts by powerful business interests employing expert witnesses to swear that black is white. Obviously the farmers still have their say while the legislation is being prepared, as do other pressure groups on both sides, but more of the argument is done before the law is in force, in a more thoroughly democratic manner. Under the terms of their 1978 Animal Protection Law, the Swiss became the first nation to commit themselves to abolition of battery cages. According to regulations issued in 1981 (which also abolished tethering pigs by the neck), Swiss poultry farmers were given ten years to achieve a total transition to less intensive forms of egg production. More recently The Netherlands have embarked on a programme of research with the aim of banning battery egg farming in favour of less intensive methods, and in 1988 Sweden passed the most comprehensive law to date for the protection of animals. By the time this is fully implemented, cattle will be entitled to graze in the open air, cows and pigs must have access to straw in stalls and boxes, sows must no longer be tethered, battery cages will be abolished, and new humane slaughter regulations will be enforced. The Swedes are also planning strict controls on the use of genetic engineering and growth hormones, and the compulsory marking of cosmetic products which have been tested on animals. If policies on battery egg farming are used as an indicator for animal welfare legislation, Norway and Denmark also score higher than Britain, while France, Belgium and the majority of other countries lag behind.[5]

One way of getting rid of animal welfare problems is to export them. This has already happened in a number of cases of relocation of vivisection laboratories in countries where controls are relatively lax; and the traditional British way of overcoming scruples about slaughtering worn-out horses was to send them across the Channel to be turned into horsemeat. The same effect can be observed in the statistics for calf exports. In the year that the phasing out of veal crates in Britain was announced, the number of calves exported to the continent rose by 81 per cent to 365,000. In the same year imports of Dutch veal, mostly for the catering trade, more than doubled.[6] Large agri-business projects will also tend to locate where planning and welfare regulations are more lax: when the West German firm Kathmann proposed the largest chicken complex in the world near Mons in Belgium, it is probable that the lower level of animal welfare activity in that country played a part in their plan to locate the 3½ million bird plant outside German territory.[7]

Farmers rightly point out the distortion of trade which arises if neighbouring countries adopt radically different welfare standards, thereby placing the farmers of one state at a competitive disadvantage. The veal industry in Britain was relatively insignificant, which made banning crates easier; also the competitive edge of crated versus 'welfare' veal is quite small. In Switzerland, labour costs and overheads were so high that home produced eggs were already more expensive than imports, which should make the transition away from battery cages less of a problem than if the Swiss had been net exporters of eggs. And in Holland, where about 70 per cent of eggs are exported, the move to less intensive systems must be seen against the background of government efforts to cut total livestock numbers before the whole country disappears beneath a sea of slurry. The Dutch already have considerable experience of alternative systems, with about 15 per cent of home egg consumption coming from deep litter birds; the shop price of these *scharrel eiren* or eggs from scratching hens is only 10 to 20 per cent more than that charged for battery eggs.[8]

The European Community's interest in farm animal welfare is largely concerned with the distortion of competition resulting when member states have different welfare standards. In 1978 the European Commission adopted as a basis for the formulation of welfare directives the European Convention for the Protection of Animals kept for Farming Purposes, a document whose principles have been agreed by the majority of European nations, both within and outside the European Community.[9] The European Convention states, among other things, that,

> Animals shall be housed and provided with food, water and care in a manner which – having regard to their species and to their degree of development, adaptation and domestication – is appropriate to their physiological and ethological needs in accordance with established experience and scientific knowledge ... [and that] The freedom of movement appropriate to an animal, having regard to its species and in accordance with established experience and scientific knowledge, shall not be restricted in such a manner as to cause it unnecessary suffering or injury.[10]

EC legislation is formulated as directives (or more detailed regulations) by the European Commission, the permanent executive body of the Economic Community. Once these have been given the unanimous approval of the Council of Ministers, they are legally binding on all member states. In the case of directives concerning farm animals, the Council of Ministers is composed of the Agriculture Ministers from each EC country. The European Parliament has no direct legislative powers, but can influence the Commission and the Council by submitting reports on particular topics, and by suggesting amendments to directives as they are being prepared. The Council of Ministers had already issued a directive on the stunning of animals before slaughter, and two on the protection of animals during international transport, when in 1979 Germany asked the Commission to consider the welfare requirements of laying hens. This followed the Frankfurt court ruling that battery cages are cruel, and there is little doubt that Germany would have welcomed the opportunity to ban battery egg production if the other EC countries had agreed to do likewise. However, it soon became apparent that there was no chance of this happening, and the Council of Ministers amended their brief to the Commission, charging them to produce a directive, not on the protection of egg laying hens, but on the protection of egg laying hens *in cages*.[11] For the next five years endless committees debated the best size and shape of battery cages, and in 1986 a directive was eventually issued, laying down minimum standards that were marginally better than the worst cages in use at the time.[12] The 450 sq cm per bird gives each laying hen a space about 20 per cent larger than a single page of this book, or approximately a quarter of what she would need to flap her wings. The need for unanimity among the ministers of the EC countries is an effective brake to radical reforms; however, it must be conceded that while the threat of a Greek veto may have thwarted plans for the eventual phasing out of batteries, even the minimum standards achieved are an improvement over existing controls in France and the southern European countries. And while Britain, for example, is adopting the requirements of the directive as it stands (except for cages containing fewer than three birds), Germany has set its standards slightly higher, while Denmark has stuck to its plans for a 600 sq cm minimum space allowance, and The Netherlands are still working towards a total ban on cages.

Although the lack of progress to date on battery cages has been disappointing for animal welfarists, the European Community's insistence on uniform standards and its ability to exclude overseas competition may yet result in gradual improvements in the treatment of farm animals. Without the combined economic power of the Community countries, such difficult moves as resisting the drug companies' pressure to allow hormones in meat and milk would almost certainly have been impossible for any single member state to have undertaken. Since Britain's accession to the Treaty of Rome, the RSPCA has played an invaluable role in encouraging Community-wide initiatives on animal welfare. In 1980 the RSPCA was instrumental in founding Eurogroup for Animal Welfare, a federation of leading national welfare organizations with the specific aim of lobbying the European Parliament and Commission.

The European Parliament, despite its lack of direct power, has taken a positive interest in animal welfare which may eventually lead to significant reforms. Richard Ryder attributes this in part to the 1982 seals campaign by the International Fund for Animal Welfare, which focused the attentions of a sympathetic public on the European Parliament, and resulted in the Community banning seal skin imports in 1983.[13] Efforts to obtain agreement on the more controversial topics of *foie gras* production, the import of frogs' legs, and bullfighting have been less successful; but in 1987 the European Parliament gave its overwhelming support to a report on farm animal welfare policy prepared by the Committee on Agriculture, Fisheries and Food under the chairmanship of Richard Simmonds, Member of the European Parliament for Wight and Hampshire East.[14] This report proposed the banning of veal crates, of tethers for sows, and the phasing out of battery cages. It also included measures for proper enforcement of the existing directive on transport of animals, and a call for the full implementation of the European Convention for the protection of farm animals in all member states. In view of the fact that the report was passed by 150 votes to nil, with two abstentions, the European Parliament can be expected to maintain pressure for its translation into Community law. Already, the Community has ratified the European Convention, but the draft directive on veal calves is disappointing, although it represents an improvement on current practice in many countries. The proposed EC regulations for pigs are also a compromise which will satisfy few animal welfarists, most of whom fear that 'creative' interpretation of the Simmonds report by interminable committees will result in legislation so watered down as to be unrecognizable. Only when the Council of Ministers can bring themselves to endorse something better than the lowest common denominator, or when the European Parliament gains more direct democratic power, will substantial legislative progress result.

A problem the European Community is already having to face is that it is one thing to make regulations or issue directives to governments to make their laws in compliance with European standards, but it is quite another thing to enforce these decisions. Despite the existence of two directives on the international transport of animals, exhaustive monitoring by the RSPCA Special Investigations and Operations Department showed that Community law was being flouted on a massive scale. The RSPCA followed 137 truck loads of live food animals, and observed that half the journeys lasted over twenty-four hours, with some taking forty-eight hours or more, and one lasting sixty-six hours. Only *one* driver stopped to feed and water the animals as required.[15] The European Commission have decided to start proceedings against France and the United Kingdom at the European Court of Justice, but there can be little doubt that in the absence of an international inspectorate with adequate resources, the law-breaking will continue.

The Community laws on slaughterhouse operations, and in particular the directive on stunning of animals before slaughter, are in equal danger of falling into disrepute. Poorer countries like Greece and Turkey lack the means to enforce these laws even if they had the will, while Spain is

Despite efforts towards standardization among EC countries, attitudes towards animals and the effectiveness of animal protection laws differ widely. The pigs in this Greek abattoir are bludgeoned to death without any form of pre-stunning. (Greek Animal Welfare Fund)

now the subject of investigation by the European Commission for deliberate disregard of the humane stunning directive. Reports in the German press have revealed that it is customary for matadors to practice their killing techniques in Spanish slaughterhouses: 'He slaughtered 40 cows that morning with the 'descabello' – a quick sword thrust into the neck. Sometimes he had to strike 3 or 4 times before the cow, afraid of death and screaming with pain, collapsed.'[16]

The traditional Iberian fiestas are the occasion for further cruelty, as pigs and lambs are killed in the streets. From the Liga Portugesa dos Direitos do Animal, Ana Isabel Pinto writes: 'They cut their throats while they are alive and let them bleed to death. Those 'parties' are supported by priests and policemen. We have a sanitary law against it, but no one executes it'.[17] It is difficult to imagine how the Commission will deal with such cases effectively, particularly when the matters involved can hardly be described as major factors in unbalancing international trade.

Trade between member states is the European Community's *raison d'être*, and there are fears among welfare organizations that trade in live animals is about to increase dramatically. In recent proposals for legislation the European Commission has set its face firmly against the spirit of the European Convention for the protection of animals during international transport, by over-riding existing British export regulations and substituting rules of a lower standard. Describing health and welfare checks at UK ports as 'barriers to trade', the Commission intend to abolish the present requirements for resting, feeding and watering animals near the port before they are sent across the Channel, and to make Britain revoke the Horses and Ponies (Minimum Values) Order which at present restricts the export of live ponies for meat. The French Rural Code for transportation, which sets a journey limit of twelve hours will also be abolished as illegal under European regulations which stipulate a maximum journey time of twenty-four hours, or longer if the animals are rested at staging points along the route. Compassion in World Farming predict a boom in exports of British lamb to Spain and Italy, where a load of 600 lambs might sell for £8 a head more than at a UK abattoir, leaving the farmer nearly £2000 better off even after paying the transport costs.[18]

This is the other side of the 'unfair competition' coin, forcing Community members to reduce existing welfare standards to a common level, rather than setting minimum obligations which the more enlightened nations are entitled to exceed. Free trade within the European Community for sentient animals is not the same as free trade in meat: as Peter Roberts points out, 'Meat doesn't suffer'. If the British Veterinary Association's policy of slaughter as near as possible to the point of production were to be adopted, there are few areas where animals would have to travel for more than three or four hours to slaughter.

The differing attitudes towards animal welfare within the European Community present a major challenge, which Britain and the other predominantly protestant nations of the north have a particular responsibility to meet. But the standards of conduct towards farm animals within the Community compare favourably with those in most of the rest of the world; particularly in the regulations governing transport and slaughter. In India, for example, cattle slaughter is primitive in the extreme despite Hindu veneration of the cow. The government is anxious to modernize, but many people are worried that improved slaughterhouses would be accompanied by an increase in the number of cattle being bred specially for meat.

Surely the worst journey in the world for farm animals today is the sheep run from Australia to the Gulf. Seven million sheep a year are exported live from Australia to the Middle East, and in the past five years more than a million have died in transit. An estimated 400,000 sheep have been turned away from Saudi Arabian ports to be hawked around the other Gulf states. According to the Saudi Agricultural Ministry, one shipload of 68,000 sheep which was rejected at Damman later turned up in Kuwait with only 21,000 animals on board; and there are repeated allegations of cruel handling, including throwing casualties overboard alive to feed the sharks. A select committee of the Australian Senate have recommended that live exports be replaced by a refrigerated carcass trade – particularly in view of the fact that when it gets to the Arab states most of the meat is frozen immediately after slaughter.[19]

With any increase in international trade there will be more problems over differences in cultural and religious attitudes to the handling and slaughter of meat animals. As in the case of the European Community, these differences may mean an erosion of standards in the countries where they are highest at present; but with increasing contact it is to be hoped that there may be an improvement in the treatment of animals in some other areas. Already a number of influential Moslems have accepted pre-slaughter stunning as being more in accordance with the spirit, if not the letter, of Islam; while in Japan there has recently been an upsurge of interest in animal welfare, with Animal Rights candidates standing in parliamentary elections and the foundation of several new humane societies.[20,21] It should not be assumed, as is too easy, that we have nothing to learn from other cultures. After all, the West may have devised the humane killer, but we also invented battery cages and vivisection. Perhaps it would be better to remember Ghandi's famous reply, when asked what he thought of Western civilization: 'I think it would be a very good idea'.

17

Little by Little

I have not told the whole story. An objective account of the changes which have taken place in farming would have paid more attention to the epidemics of food-borne disease, the horrific cruelty to animals and the massive environmental damage that occurred in the days before we were lucky enough to have gold-top milk and battery eggs. But the food and farming industry spends millions to advertise the boons and blessings they shower on us, and if we seek complacency they will offer it in abundance. So the picture I have tried to paint unashamedly highlights areas where reform is needed, and the historical comparisons drawn have been intended to show that not all current practices are the best, rather than to point back to a golden age when everything in the farmyard was lovely. In some areas matters are actually getting worse; in many others progress has been made but things could still be a lot better. Many reformers cherish the vision of radical change, many if not most animal welfarists are vegetarians or vegans, but those who are also pragmatists recognize that there are numerous small improvements which can more realistically be aimed for.

Improved food labelling would be a relatively cheap step which could do much to improve the quality of food, and enable the shopping public to make more informed choices from the health angle, and also between alternative husbandry methods. The labelling of additives with E numbers had a huge impact on public awareness; so could compulsory information on saturated fats or maximum residual chemical levels. Meat which had been produced using growth promoters or pre-slaughter tenderizing agents could be labelled accordingly, and the dyestuffs used to colour farmed salmon and battery egg yolks could be declared. Only those with an inferior product have anything to fear from stringent labelling regulations; the extreme example, as already mentioned, is that of top-class wine producers who benefit immensely from the tight controls exerted by their industry.

A necessary accompaniment to improved labelling would be an

expansion of quality testing by consumer protection authorities to avoid fraud. The supermarket chains already have extensive quality control facilities, and jealously strive to maintain spotless reputations, but when spot checks reveal up to 25 per cent of butchers' shops cheating trading standards, the need for more rigorous enforcement becomes apparent.[1] In the case of tests for residues, the techniques are often complex and expensive, but it is remarkable how quickly tests have been perfected to analyse meat for hormone residues, etc. Currently tests are being developed to identify whether or not food has been irradiated; and in Switzerland battery eggs masquerading as free range can be detected using an ultra-violet technique which picks up marks left on the egg shell by the cage wires.

The European Community has begun to recognize that the consumer may have an interest in the welfare aspects of food production. In 1985 the Community introduced regulations under which eggs are labelled as free range, deep litter or perchery eggs according to the method of production. These systems are defined in the regulations, which are intended to be in the interests of consumer choice.[2] Regrettably, the minimum standards for free range have been set too low: the hen-houses permitted by the European Community are so large that some birds never go outside at all, and the allowance of 400 birds per acre of range is so high that it is difficult to maintain a healthy colony in the British climate. Battery eggs are excluded altogether from the labelling requirements, which have manifestly failed to inform the public effectively. In 1988 a poll revealed that 34 per cent of shoppers still thought that eggs labelled as 'Farm Fresh Eggs' were free range.[3] In the United Kingdom, the government has refused to insist that ritually slaughtered meat sold to the general public should be so labelled – though to be fair such a requirement would be extremely difficult to enforce.

Although it may never be possible to devise a test which would show how a particular piece of butcher's meat met its end – and this is one argument for a total ban on ritual slaughter by inhumane methods – there is tremendous scope for the introduction of controls which would at least enable individual carcasses to be traced back to their source. At present, with overcrowded markets, some beasts changing hands several times, and the possibility of confusion in busy slaughterhouses, there is very little chance of a retailer tracing the origin of an unsatisfactory carcass. This is particularly unfortunate, since under British law he is responsible to the customer for the quality of the meat, even if it should prove to be contaminated by a cause completely beyond his control. In Holland, quality control in the pig industry is being transformed by a law which makes it compulsory for all pigs to be implanted with a passive electronic transponder (costing about £1). This remains in the carcass after slaughter, and enables the producers of diseased or poor quality carcasses to be readily identified.[4]

Improved labelling could do a lot for food quality, but where there are serious doubts about the safety of processes or additives, unless there is a genuine demand – from the consumer (who bears the risk) rather than from the producer – then it will be more satisfactory to ban them. In some cases separate labelling is not practical, so what is not generally acceptable must be banned altogether. The EC ban on hormones is a case in point, though it also illustrates the need for effective policing: in 1988 thousands of German veal calves were impounded and slaughtered after it was discovered they had been illegally implanted with steroids, and £250,000 worth of similar drugs were seized in Ireland before they could be sold on the black market.[5] Another example of a product where labelling is not an option is milk. The milk from many farms is mixed at creameries, which makes it impossible to label milk from cows taking part in trials of the genetically engineered hormone BST, even if the government had been less inclined to secrecy.

Even when labelling is possible, as with phosphate-treated meat products, it can be argued that the public needs to be protected from itself: faced with two ham joints of apparently similar size, how many shoppers could resist the one which was half price, for all that it may be mainly water and phosphates? Legislative control of such practices may be denounced as paternalistic by the meat processors, but there are perfectly respectable precedents. Building control regulations, for example, protect the public from cheap and shoddy housing, which would no doubt appear very attractive pricewise.

The quality of animal products from modern farms is influenced by the animals' life style and breed characteristics, and also by their diet. Here again, there is a strong case for stricter control on food safety grounds alone; particularly regarding the diet on which the livestock are fed. The alarming incidence of *Salmonella enteritidis* in eggs, and the scandal of the mad cow disease BSE are both directly attributable to the use of feeds containing inadequately sterilized offals. Although the use of offal in cattle feed has now been banned, huge damage has been done, and given the British Ministry of Agriculture's disgraceful laxity with the feed compounders in the past, it is most unlikely that we have seen the last of trouble from this source. The problems of epidemic disease and mass cross-infection endemic in modern stock farming and meat processing have already been discussed, and it is clear that the whole area is one where tighter controls are essential. It is no use bleating that the housewife or restaurateur should be more careful, or should give up making their own mayonnaise: surely we are quite entitled to expect that the food we buy is at least as free from harmful infections as it used to be in the past.

Stringent controls should be applied at all stages of production of the basic foods such as chicken, eggs, milk and so on. But just as Japanese gourmets are entitled to dice with death as they tuck in to their prized blow-fish, so should the specialist consumer in Britain be able to get unpasteurized milk and cheese, and other 'farm gate' products which command high prices on grounds of their superior flavour. If such luxury goods are dangerous to eat (which is far from proven), they could carry a clear warning, as cigarettes do now. This policy has in fact been adopted by the Ministry of Agriculture for the sale of unpasteurized 'green top' milk at farms and dairies. In general, though, the efforts of the Ministry to date regarding food quality have been depressing in the extreme. They have found it only too easy to clamp down on 'soft' targets like the producers of unpasteurized cheeses or free range eggs, while leaving the 10 million birds per year broiler plant unscathed, and our disgusting public slaughterhouses underfunded and undersupervised. With this record, can we reasonably hope that irradiation will solve all our problems, or will it merely prove another new menace to public health?

If food labelling cannot wholly protect public health, then neither can it reliably save the public conscience. Again and again opinion polls show the mass of the public to be revolted by the excesses of factory farming, but only a minority seek out and pay for free range produce. Cynics might argue that this is evidence that for most of us our concern is only skin deep, but this is not necessarily true. Suppose the abolition of the Slave Trade had involved the establishment of a sugar industry based on free labour, and had had to wait until the public stopped buying the cheaper product of the slave plantations; we should very probably still be waiting for its demise, which was brought about at a stroke once public outrage was translated into legislative practice. Legislation could quickly kill the battery egg industry, provided it was effective throughout the European Community, and tariffs were maintained against battery imports. Similarly, only legislation will prevent the suffering of farm animals taken on endless journeys to market for the sake of a few extra pounds. Free Trade is an excellent idea in theory, but sentient animals, like slaves, demand an exception.

The proponents of established factory farming systems, like the slavers of old, frequently talk as if the world would end if their favoured methods were abolished. It is fair enough to expect reformers to give some idea of what system they would use instead, but to expect every last detail of cost and practical operation to be made clear in advance is surely excessive. Nothing concentrates the mind like necessity, and it is notable that progress on alternative egg production methods has been remarkably rapid in countries such as Switzerland and Holland, where the phasing out of batteries has been seriously implemented. The practical alternatives, apart from free range, are deep-litter sheds which are relatively simple barns with soft litter, a dung-pit and nest boxes; or percheries in which the hens are still free to move around, but much higher stocking densities are achieved using elaborate arrays of wire perches. The details of these methods have been discussed elsewhere; the added production costs over the battery system are of the region 5 to 20 per cent.[6,7]

Pig farming has been less dominated by a single method than egg production, and alternatives to the highly intensive system of permanent tethering have therefore evolved naturally. Outdoor pig rearing can still be economic in some situations, and the most promising development undercover is the introduction of electronically operated feeding stations whereby each pig's diet can be individually monitored and controlled. Likewise, alternative methods of veal production have been shown to be competitive with the use of crates, and so there should be no major economic argument against a change to the more humane system.

If farmers were obliged to provide their stock with comfortable bedding, a great improvement in general welfare would result. The natural nesting behaviour of hens and pigs is frustrated if they have no access to straw or similar materials; while the lack of bedding for cows is probably the greatest single welfare problem of the modern dairy industry. Feeding cattle on silage makes them excrete more water, and with the increasing cost of straw farmers have tended to move away from traditional straw bedding, with the result that lameness and environmental mastitis have become much more common. For cows and pigs, methods are now available which reduce the amount of straw used, but without involving the extreme discomfort of plain or slatted concrete floors. Free housing with cubicles for resting requires about a quarter of a ton of straw per cow for the winter's bedding, compared with a ton each in a simple yard; cow mats can maintain comfort while reducing straw use still further.[8] The 'strawflow' system for growing pigs uses a sloping floor which means no cleaning is required, and avoids the slurry problem associated with hard flooring, with the use of about a quarter of the normal amount of straw.[9]

The preface to the official Farm Animal Welfare Council codes of practice lists ten basic provisions which should be made for farm animals:

comfort and shelter;
readily accessible fresh water and a diet to maintain the animals in full health and vigour;
freedom of movement;
the company of other animals, particularly of like kind;
the opportunity to exercise most normal patterns of behaviour;
light during the hours of daylight, and lighting readily available to enable the animals to be inspected at any time;
flooring which neither harms the animals, nor causes undue strain;
the prevention, or rapid diagnosis and treatment, of vice, injury, parasitic infestation and disease;
the avoidance of unnecessary mutilation; and
emergency arrangements to cover outbreaks of fire, the breakdown of essential mechanical services and the disruption of supplies.[10]

The codes themselves give excellent advice on how these aims could be achieved, emphasizing in particular the need for a high standard of stockmanship, but they are advisory rather than mandatory. Very few farmers follow the codes; if they do, for example, give 'consideration' to installing fire alarm systems, they presumably consider the expense is not worth while. If the provisions of the Codes of Recommendations were made law, and enforced on the farm, in transit, and at markets and slaughterhouses, we should have a livestock industry we could really be proud of. This requires courageous legislating instead of wimpishly toeing the line of minimum EC standards; and an improvement in funding and operation of the Ministry of Agriculture inspectorate. At present their visits to farms are too infrequent, and the farmers are given advance notice of their calls, which rather spoils the point.

The question of stockmanship is crucial to farm animal welfare, and to the farm's profitability. Unfortunately, as farms get larger and more mechanized the ratio of humans to animals drops, and the number of 'contact hours' each animal has with the stock worker declines. The constraint of animals in intensive units is partly a response to this lack of individual care: as pig farms expanded rapidly many farmers lacked the time or ability to supervise large numbers of loose housed sows, so they took to tying them up permanently. The Welfare of Livestock (Intensive Units) Regulations 1978 state that livestock must be 'thoroughly inspected' by a stock keeper not less than once per day. As Chickens' Lib point out, even if he spent the whole day inspecting stock, this would only allow half a second for each bird in a normal poultry unit – hardly enough for a thorough inspection.[11] However well designed large livestock units are, the sheer numbers involved make it impossible to care for the animals as individuals. Some vets are worried at being forced to take a less caring attitude about the individual animal: 'We have had the difficult job thrust upon us of attempting to advise at herd or flock level when our teaching and very often our instincts are directed to the individual'.[12] The only solution to this problem – which is most unlikely to be adopted in the United Kingdom – is to do as the Swiss have done, and put statutory limits on the numbers of animals which can be kept in one flock.

Ever-increasing flock sizes, and the concentration of intensive farms in certain geographical regions, have already caused such pollution that Denmark and Holland have been forced to prohibit further growth of livestock numbers. The British Government should take steps to ensure that factory farmers pay the full cost of any pollution and environmental damage they cause, that they contribute to local services by paying rates on farm buildings, and that farm developments are fully controlled by democratically accountable local planning committees. Similar rules should be imposed on fish farms, and the privileged position of the Crown Estate Commissioners abolished. The release of genetically engineered organisms into the environment should be prohibited, at least until full international agreement is obtained.

It would be churlish to concentrate too much on the negative impact of farming on the countryside. Traditional farming practice has shaped the landscape, and there is no reason why present day farmers cannot make aesthetically pleasing contributions to our heritage. However insolent the claim made by Buitelaar that their illegal earthworks in Lincolnshire improved the landscape, and 'could in time qualify for registration as an ancient monument', there are many legitimate farming and forestry works for which these claims could quite justly be made.[13]

With the need to cut back EC production of beef, milk and wheat the government has an ideal opportunity to encourage more socially desirable long-term farming policies. Rather than simply setting aside productive farmland or releasing more surplus acres for housing and industrial development, lower input farming of crops and the return to the land of some of our permanently housed livestock would be policies which would both make the countryside more attractive now and conserve soil fertility for generations to come. Even without government assistance, organic farms are already being run profitably by enthusiasts, and with more encouragement from Whitehall the number of such enterprises could be greatly increased.[14] Much credit must go to the Soil Association, whose pioneering efforts and determination to maintain strict standards have prepared the way for the development of organic farming in Britain. True to the spirit of the organic movement, the Soil Association insists on humane standards of animal husbandry, as well as strictly limiting chemicals used in feeding or for therapeutic purposes.[15] There is inevitably a slight risk of enthusiasts withholding helpful medication in the interests of maintaining organic purity, but current reports indicate that animal health on organic farms is so much better that this is not likely to be much of a problem for anyone except the pharmaceutical industry. In any case, the Soil Association standards make sensible provision for sick animals to receive necessary veterinary treatment.

Independent of the organic movement, but complementary to it, is the British Government's new policy of designating environmentally sensitive areas, in which farmers are offered payments under five-year contracts to farm in ways which conserve the landscape, habitat or historic value of their farms. This is an infinitely better arrangement than the system of compensation provided under the 1981 Wildlife and Countryside Act, which is little more than a rip-off at the taxpayer's expense (see pages 193–4). The environmentally sensitive area scheme deserves to be extended, and should help reduce overproduction. It seems a particularly helpful way of making social payments to farmers in difficult areas; far more so than the present subsidy system which encourages gross overgrazing and environmental degradation.

As one of the more sensible pieces of legislation to emerge from the Ministry of Agriculture since the Second World War, the environmentally sensitive areas scheme deserves to be welcomed. But there are many people who feel that the Ministry is incapable of the rapid changes that are now urgently required, and should be abolished, either to be replaced by a more consumer-oriented Ministry of Food, or simply to be carved up between the existing Trade, Environment and Health departments. At the van of the Ministry's critics is Sir Richard Body, a former chairman of the Commons Select Committee on Agriculture; and unless the Ministry can get itself out of the pocket of the agri-business lobby and show some more concern for the consumer, the small farmer, and the countryside, his case is unanswerable.

Many of these proposed reforms need legislation; in other cases, such as the need to improve slaughtering conditions, an increased commitment to enforcing existing regulations would go a long way. But none of them would have even been advocated if it were not for the dedication and concern of individuals such as Ruth Harrison and Peter Roberts in mobilizing public opinion in sympathy with the 'animal machines' of modern farming. An increasingly sympathetic press, fed with information by the RSPCA, Compassion in World Farming, Chickens' Lib and many smaller charities, has played a significant part in getting the message across to the consumer. And consumer power, recognized by small independent farmers and by supermarkets, has demanded alternatives and proved that it is possible to provide them. The major chains have recognized the fashionable appeal of 'green' products, and are competing with each other to find adequate sources of supply; and a new country-wide network of organic food stores is also being planned.

One of Compassion in World Farming's slogans is 'Don't buy your food in ignorance'. As consumers, our individual choices may seem of little effect, but their aggregate contribution may yet be a powerful force for change. Now that more and more supermarkets offer a choice, we can only hope that a growing number of shoppers will have the knowledge, and the inclination, to exercise that choice in favour of a more humane and sustainable kind of agriculture.

Appendix 1

What to Buy – Food Quality

Produce	Main dangers	What to look for
Game	Lead, radioactivity.	Avoid venison, quail, etc., unless definitely wild.
Lamb	Radioactivity, sheep dip residues. Indoor fattened lamb is tasteless.	Look for Scottish/Welsh lamb from hill sheep for best flavour.
Beef, veal	Illegal hormones, or foul taste if fed intensively.	Look for Angus/ Hereford beef from organic growers.
Fish	Heavy metals. Farmed fish tasteless and flabby.	Avoid salmon and trout unless definitely wild.
Dairy	Pesticides, heavy metals, PCBs and other residues.	Look for unpasteurized cheeses for best taste. Drink semi-skimmed to beat most of the residues.
Eggs	Some risk of salmonella. Fishy (or worse) taste.	Look for free range – and check they are fresh.
Pork	Drug residues, salmonella, generally tasteless.	Look for outdoor herds – preferably organic.
Ham, sausages	Residues, phosphates and other additives.	Read the label!
Chicken (and turkey)	Covered in food-poisoning bacteria. Foul taste and texture.	Look for free range birds from organic growers.

←Worse——Food quality——Better→

Appendix 2

What to Buy – Animal Welfare

Produce	Welfare aspects	What to look for
Wild game/fish	Lead natural lives, death may be painful and slow.	Check it really is wild.
Dairy produce	Cattle stop giving milk if *too* badly treated. Intensive feeding and permanent housing are increasing stresses.	Not much choice at present given the Milk Marketing Board's stranglehold.
Farmed fish and game (venison)	Unnatural confinement, and stress at slaughter. Even fish can suffer!	Avoid quail, rabbits and other small animals from battery cages.
Lamb	Often neglected and underfed otherwise fairly natural.	Try to buy local meat, to reduce live transport.
Beef	Often grown under cover, on dirty slatted floors.	Local or organic beef probably best.
Chicken (and turkey, duck)	Crowded, and not cared for individually. Suffer many complaints of overbreeding.	Look for free range birds. Geese are kept outside at least some of the time.
Pork, ham, bacon	Many sows are permanently tied up, and young pigs are often very overcrowded.	Look for meat from free range herds, eg. Heal Farm, The Real Meat Company.

↑ Better — Worse ↓

Produce	Welfare aspects	What to look for
Eggs	Battery hens can hardly move – how would you like it?	Avoid eggs that aren't labelled free range (or deep litter, perchery).
Veal	White veal is still from calves living in dark crates.	Look for British, 'welfare' veal, or dark pink colour.
Foie gras	Many birds die rather than suffer the abuse of *gavage*.	Avoid at all costs.

Animal welfare: ←Worse——Better→

Appendix 3

What to Buy – Environmentally

Game	Game management may be ecologically harmful, particularly towards predators, but it is less disruptive of the environment than other types of farming.	↑ Better
Lamb	Sheep graze otherwise unproductive areas, but cause erosion and devastate vegetation.	
Wild fish	Trawling damages sea-bed; lost gear snares seals, otters, etc.; unintentional by-catches of other species; overfishing can starve out other wildlife.	
Dairy produce	Traditionally sustainable, but intensive methods cause water pollution and degrade pasture land.	
Beef	Intensive beef rearing wastes cereal fodder and often causes pollution. Beef cattle for hamburgers follow the burning of tropical rainforests.	
Eggs, chicken, other poultry	Intensive poultry eat high protein feed such as fish meal, are very smelly, and their waste highly infective.	
Pork, bacon, ham	Inefficient users of fodder when kept intensively. Pigs produce vast amounts of contaminated slurry which is a major disposal problem.	
Farmed fish	Every farmed fish has eaten five times its own weight of wild fish. The wasted feed, excrement and pollution by numerous chemicals pose a major new threat to rivers and coastal waters.	Worse ↓

Appendix 4

Environmental Contamination of Animal Produce

These chemicals are ingested by grazing animals, or are present in crops or animal products used in feeds. They are not specifically connected with intensive livestock husbandry.

Contaminant	Sources	Effects
Heavy Metals (mercury, lead, cadmium, arsenic)	Industrial pollution, exhaust fumes, domestic waste. Lead shot in game. High levels of mercury and arsenic in fish.[1]	Cumulative poisons of liver and kidneys. Lead causes lethargy, anaemia and damage to central nervous system.[1]
Pesticides		
Organochlorine compounds: DDT, dieldrin, aldrin, HCH (lindane)	Mostly banned now in the United Kingdom, but still present in the environment in declining quantities. May occur in some imported meat or dairy produce.[2]	Very persistent. Acute toxicity low, but can accumulate in fatty tissues and eventually cause cancer or birth defects.[3]
Organophosphorus compounds and pyrethrins	Widely used on growing plants and to preserve crops after harvest.	Much less persistent, not found as residues in animal produce.[2]
Arsenicals	Could be used again if more modern compounds banned.	Acutely toxic, main danger accidental overdoses.
Dioxins	By-products of pesticide manufacture.	Persistent, very toxic, mutagenic, teratogenic.[4]

Contaminant	Sources	Effects
PCBs	Electrical transformers and many other industrial uses. Concentrate in milk and fats, especially fish fat.	Similar to organochlorine pesticides.[3]
Radionuclides	Fall-out continues from bomb tests in 1960s and more recently. Nuclear reactor accidents, reprocessing facilities.	Radiation dose is low compared to background, but Caesium 137, Tritium and Strontium 90 readily absorbed in the body, and may cause cancer.[3,4]

Another group of contaminants which occur naturally under certain circumstances, and are sometimes present in animal feeds, are

Contaminant	Sources	Effects
Mycotoxins Aflatoxins, Ochratoxins	Formed by moulds, especially on cereals, oilseeds and nuts. Aflatoxins formed mainly in tropics; ochratoxins in temperate climates too.[5]	Acute toxic properties and very carcinogenic.[3] (May be less dangerous in animal flesh than if ingested directly; p. 81)

Appendix 5

Contamination from Animal Treatment

The majority of these chemicals are more likely to be found in animals from intensive husbandry systems. The largest residues come from illegal use of drugs, or when withdrawal times are not observed before slaughter.

Contaminant	Use	Effects
Antibiotics (based on micro-organisms)		
Feed additives, e.g. monensin, virginiamycin, bambermycin	Growth promoters (monensin also used as a coccidiostat for poultry). Not used in human medicine.	Detectable residues seldom found, as these substances are not absorbed from the animals' gut.[6]
Penicillins	Cheap general purpose antibiotics. Can be given in feed or by injection.	Residues not very toxic; possibility of allergic reaction in some people.[7]
Aminoglycocides, e.g. streptomycin, gentamicin	Treatment of mastitis and other infections. Must be injected.	Accumulate in kidneys; can affect hearing or heart function.[8]
Tetracyclines	Very broad-spectrum, popular veterinary antibiotic. Of particular value for respiratory infections.	Not exceptionally toxic.[3]
Chloramphenicol	Powerful antibiotic, used to treat pneumonia, peritonitis, etc. Not licensed for animals in the United States of America.	Suspected of causing leukaemia, even at very low doses.

Contaminant	Use	Effects
Antibacterials (synthetic)		
Feed additives, e.g. sulphadimidine	Growth promoters, especially for pigs. Sulphonamides also used therapeutically for infections such as rhinitis.	Very common residues, as withdrawal period often not observed. Suspected carcinogen.[9]
Nitrofuranes, e.g. furazolidine	Prevention of enteritis in pigs, poultry and calves.	Carcinogenic and mutagenic.[3]
Carbadox (R)	Growth promoter and antibacterial for pigs.	Carcinogenic and mutagenic.[3]

The synthetic antibacterials are generally excreted much more slowly than the antibiotics, and require withdrawal periods of up to four weeks. Effects listed relate to toxicity, not to the additional danger of the spread of antibiotic resistant strains of bacteria (especially salmonella species).

Contaminant	Use	Effects
Anabolics (hormone growth promoters)		
Natural steroids, e.g. testosterone, progesterone	Use forbidden in the European Community, but abuses difficult to detect (present naturally in meat).	All these compounds act as sex hormones. Residues may occur locally at high levels near the implant site. DES is carcinogenic. Butchers may be specially at risk (p. 77)
Synthetic sex hormones, e.g. DES, Zeranol	Replicate the action of the naturally occurring steroid hormones.	
BST (genetically engineered bovine somatotropin)	Regular injections prolong lactation peak in milk cows. Not yet permitted in the European Community, despite strenuous lobbying by manufacturers.	Toxic effects of residues still unestablished.
Beta-Agonists	Latest growth promoting drugs. Not yet allowed in the European Community, but in use illegally.	Toxic effects of residues still unestablished.

Contaminant	Use	Effects
Tranquillizers e.g. Azaperone	Given before transport or slaughter (particularly to pigs), to avoid meat spoilage due to stress.	Residues may occur near point of injection. Not particularly toxic.[3]
Parasiticides		
Anthelmintics	Worming compounds. Mainly used for cattle.	Affect liver. Seldom found as residues.[9]
Ecto-parasiticides	Insecticides such as lindane used for sheep dips. Now generally replaced by less persistent compounds. Aquagard (R) louse treatment for farmed salmon.	See appendix 4. Lindane still occurs in some UK and EC lamb, and DDT in some New Zealand products.[2]
Larvadex	Fed to laying hens to kill insects in their droppings.	Teratogenic.[10]
Other chemicals		
Arsenic compounds	Used to improve feathering and egg bloom.	Cumulative toxin, but residue levels higher in many natural products.[1]
Mineral supplements	To correct deficiencies in diet.	No toxic effect at normal doses.
Carotenes	Colour egg yolks and flesh of farmed fish.	Impair vision.[11]
Probiotics	Encourage gut bacteria and improve growth rate.	No known toxic effect.
Papain	Injected into cattle before slaughter as tenderizer.	No known toxic effect.

The practice of recycling offals, corpses and faeces can lead to unintentional accumulation of heavy metals or antibiotics in the animals' diet. General use of medicated feeds has given rise to several incidents in which use of the wrong antibiotic, or accidental cross-contamination at the mill, has adversely affected the livestock.[12]

Appendix 6

Contamination from Meat Processing

Mass-produced meat products contain a number of chemicals, and traditional curing methods have also been criticized for the incorporation of substances of doubtful safety.

Contaminant	Process	Effects
Polyphosphates	Used in re-forming meat, and to retain water, making products heavier.	Not generally believed to be toxic.
Nitrites	Traditional curing agent, kills bacteria, including *Clostridium botulinum*	Carcinogenic nitrosamines formed during curing or cooking.[13]
PAHs (polyaromatic hydrocarbons)	Produced during smoking and grilling.	Some PAHs are powerful carcinogens.[2]
Sulphites	Preservative in sausages.	May cause asthma.[14]
Colourants	Remedy bleaching effect of sulphites.	There are too many food colours and additives to begin to list their toxic properties here.[15]
Other additives	Emulsifiers, anti-oxidants, flavour enhancers, etc., etc.	
Disinfectants		
Chlorine compounds	Used in washing surfaces and tools during cutting and handling of meat.	Can react with meat to produce several toxins.[16]

Contaminant	Process	Effects
Surface active compounds	As above.	Can inhibit digestion. Meat probably not a major source of these.[3]
Packaging film	Cling films widely used to display meat in shops.	Some film plasticizers are highly carcinogenic.[17]
Irradiation	Now allowed for preserving food, which has to be labelled.	Free radicals formed can cause substantial changes in chemistry. Poisonous gases and carcinogens may be produced, and vitamins are destroyed.[18]

Appendix 7

Microbial Contamination of Animal Products

These zoonoses, or diseases commonly caught from animals or animal produce can also be contracted from other human beings. In the majority of cases, the origin of the infection can only be guessed at (see chapter 6).

Organism	**Origin and frequency (approx. UK cases per year)**	**Effects**
Salmonella: cause approx. 50 deaths per year		
S. typhimurium (drug sensitive)	Mainly pigs and poultry (6000)[19]	Vomiting, diarrhoea, occasional septicaemia.
S. typhimurium (drug resistant)	Beef, dairy products (600)[20]	As above. Antibiotic treatment more difficult.
S. enteritidis	Eggs and meat products (10,000)[19]	As above.
Other *Salmonellae*	Various (6000)[19]	As above.
Campylobacter	Chicken (25,000)[21]	Similar 'food poisoning' symptoms to salmonella.
Clostridium perfringens	Meat products (2000)[22]	Diarrhoea, stomach pains.
Listeria	Dairy products? (silage) (60)[19]	Abortion or neonatal death. Meningitis. Often fatal.

Many other diseases can be caught via animal foods on very rare occasions.

Appendix 8

Welfare Aspects of Intensive Farming

Broiler poultry	Lack of stimulus, causing tendency to hysteria. Crowded conditions may encourage cannibalism, and necessitate beak trimming. Average growth rates depressed through competition for feed. Overheating may occur in summer. Hock burns from damp litter. Selection for very rapid growth stresses bones and lungs to limit.
Caged laying hens	Lack of stimulus and exercise. Frustration of feeding, nesting and pre-laying behaviour. Brittle bones. Cannibalism. Mouth ulcers. Fatty degeneration of liver. Increased susceptibility to disease.
Quail	Similar problems to caged hens.
Breeding sows	Close confinement causes boredom, stereotyped behaviour. Lack of exercise leads to farrowing difficulties, and lameness from lack of wear on toes. Hard floors and bars can cause injuries. Fertility and feed conversion depressed by constant tethering. Discomfort from use of farrowing crates.
Growing pigs	Wire cages for weaners uncomfortable. Sweat boxes or other very crowded systems unsatisfactory. Overcrowding and lack of bedding lead to fighting or tail biting.
Veal calves	Lack of light and space. Inadequate diet causes anaemia. Confinement, contact with faeces and early weaning encourage disease. Maternal deprivation leads to stereotyped licking and sucking of self or others. Majority of calves from the United Kingdom face stressful transportation to continent.
Beef cattle	Discomfort from slatted floors when kept under cover. When fed outside, often penned in slippery and dirty enclosures.

Milk cows	Inadequate bedding main source of discomfort, and lying on hard dirty floors also causes mastitis. Intensive feeding causes cows to 'burn out' after only a few years.
Mink, foxes	Stereotypies indicative of boredom and frustration. Defective maternal behaviour.
Sheep, deer	Outdoor flocks exposed and often hungry. Housed sheep suffer increased stress, and more likely to be fed inappropriate diet. Live export of sheep to the European Community a major welfare problem. Deer difficult to transport and slaughter humanely.
Fish	More difficult to assess degree of suffering. Caged fish wear down dorsal fin in efforts to escape from net. High disease levels. Migratory instinct frustrated?

Appendix 9

Environmental Impacts of Livestock Production

	Pollutant	Effect
Cattle	Yard washings	De-oxygenation of watercourses, spread of resistant bacteria.
	Silage effluent	De-oxygenation and poisoning of watercourses. Odour pollution.
	Slurry spreading	Smells, dispersal of bacteria, possible run-off to watercourses or contamination of crops.
Sheep	Sheep dips	Have contributed to general environmental pollution by organochlorine compounds. New types should be less harmful.
Pigs	Slurry	As for cattle slurry. May contain high levels of drugs or copper salts. Available land frequently inadequate for safe disposal.
	Exhaust fumes from housing	Smells, bacterially contaminated dust or vapour.

	Pollutant	Effect
Poultry	Exhaust fumes from housing	As for pigs.
Fish	Faeces, waste food and general detritus	De-oxygenate water and provide nourishment for toxic algae.
	Pathogenic bacteria	Infection of nearby farmed fish or wild stocks.
	Interbreeding with wild fish (through escape or release)	Possible enfeeblement of wild stocks.
	Antibiotics	Development of resistant micro-organisms.
	Pesticides, e.g. Aquagard (R)	Toxic to crabs and lobsters.

Numerous toxic chemicals are used in animal production, which may escape into the environment during use or after improper disposal. Such escape is inevitable in fish farming. The main chemicals used in aquaculture are listed in more detail in the following appendix.

Appendix 10

Principal Chemicals used in Aquaculture

Chemical (F = can be given as feed additive)	Purpose
Antibiotics	
Macrolides (F) (erythromicin)	Bacterial kidney disease (BKD).
Tetracyclines (F)	Columnaris, enteric redmouth, furunculosis, gill disease, vibriosis. Widely used.
Trimethoprim (F)	Used to increase efficacy of sulphonamides (see below).
Antibacterials	
Sulphonamides (F) (with trimethoprim, e.g. Tribrissen, Romet 30)	BKD, columnaris, enteric redmouth, furunculosis, vibriosis.
4-Quinolones (F), e.g. oxolinic acid	Columnaris, enteric redmouth, furunculosis, vibriosis.
Nitrofuranes	No longer widely used because of mutagenic activity.
Quaternary ammonium compounds, e.g. Roccal	Surface active antibacterial used to treat columnaris, fin rot, gill disease.
Other biocides	
Malachite green	Fungicide, widely used on eggs, also as an antibacterial to treat fin rot, gill disease (not permitted in the United States of America).
Formalin	Widely used against ectoparasites, also as egg fungicide.

Chemical	Purpose
Dichlorvos (Nuvan, renamed Aquagard)	Organophosphate parasiticide, essential for control of salmon lice. Very toxic, especially to crabs and lobsters.
Disinfectants	
Halogen compounds	General use on tanks, equipment and for footbaths.
Caustic soda	Disposal of diseased fish, etc.
Anaesthetics	
MS222, benzocane	Used before stripping eggs, also before transport.
Carbon dioxide	Occasionally used before fish are killed.
Astaxanthin	Dye added to feed to colour salmon flesh pink.

The majority of these chemicals leave residues in the fish, and withdrawal times are specified before the fish can be killed for eating. Only diseases and medications relevant to the United Kingdom have been listed.[23–5]

Notes and References

Chapter 1 The Agricultural Enterprise

1 Leakey, R. E. *The Making of Mankind* (Joseph, London, 1981) p. 206.
2 Leviticus, ch. 25, v. 4.
3 Virgil. *Georgics*, Book 1, lines 77–81.
4 Pimental, D. et al. 'Pesticides, Insects in Foods and Cosmetic Standards'. *Bioscience*, 27 (1977).
5 Association of Agriculture. *Food Crops and Pesticides* (Association of Agriculture, London, 1988).
6 Pliny, translated by Holland, P. *Natural History*, Book 6 (Centaur Press, Fontwell, 1962) p. 67.
7 Gibbon, E. *Decline & Fall of the Roman Empire* (Dent, London) vol. I, p. 334.
8 Serpell, J. *In the Company of Animals* (Basil Blackwell, Oxford, 1986).
9 Leakey, R. E. *The Making of Mankind*, p. 210.
10 Virgil. *Georgics*, Book 3.
11 Each parent has a red and a white gene, so there is a 25 per cent chance of the offspring being red, a 25 per cent chance of its being white, and a 50 per cent chance of its being roan.
12 Webster, J. *New Scientist*, 21 July 1988, p. 41.
13 This has been partly overcome with broiler fowls, by using 'mini mothers' with a non-transmissible dwarfing gene, to lay eggs from which normal-sized offspring are hatched.
14 The *Independent*, 19 September 1988.
15 *AgScene*, no. 91 (1988).
16 Kiley-Worthington, M. *Behavioural Problems of Farm Animals* (Oriel Press, Stocksfield, 1977) p. 50.
17 Kiley-Worthington, M. *The Behaviour of Horses* (J. A. Allen, London, 1987) p. 102.

Chapter 2 From Farm to Factory

1 Varro. *Rerum Rusticani*, Book 3.
2 Thomas, K. *Man and the Natural World: Changing Attitudes in England 1500–1800* (Allen Lane, London, 1983) pp. 93–4.
3 Chickens' Lib. Fact sheet, no. 11, 124.
4 Haldane, A. R. B. *The Drove Roads of Scotland* (Edinburgh, 1952).
5 Blunt, W. P. (ed.). *Intensive Livestock Farming* (Heinemann, London, 1968).
6 Compassion in World Farming. Campaign poster.
7 *MAFF Statistics* (Government Statistical Service); *Statistical Abstract of the United States* (US Bureau of the Census, Washington DC, 1988); *Japan Statistical Yearbook* (1988).
8 Poultry Research Centre. *R & D Publication No. 16* (Poultry Research Station, Roslin, Midlothian, 1982).
9 *AgScene*, no. 79 (1985).
10 Hanson, B. S. 'Diseases and Egg Quality', in Carter, T. C. (ed.), *Egg Quality*. British Egg Marketing Board Symposium no. 4 (Oliver & Boyd, Edinburgh, 1968).
11 Chickens' Lib. Fact sheet, no. 23.
12 Ibid., no. 413.
13 Chickens' Lib leaflet.
14 Fox, M. *Agricide* (Schocken Books, New York, 1986) p. 4.
15 Harrison, R. *Animal Machines* (London, 1964).
16 *Poultry World*, 14 June 1984.
17 Blaxland, J. D. and Borland, E. D. 'A survey of "normal" broiler mortality in East Anglia', *Veterinary Record*, 101 (1977) pp. 224–7.
18 *MAFF Statistics* (Government Statistical Service).
19 The *Independent*, 29 December 1988.
20 *AgScene*, no. 74 (1984).
21 *Animals International* 5, no. 18 (1985).
22 *AgScene*, no. 88 (1987).
23 *Chemistry in Britain*, November 1989, p. 1075.
24 *AgScene*, no. 73 (1983).

Chapter 3 Taming and Ownership

1 Quoted by Thomas, K. *Man and the Natural World: Changing Attitudes in England 1500–1800* (Allen Lane, London, 1983) p. 171.
2 *Animals International*, Winter, 1983.
3 *AgScene*, no. 90 (1987).
4 Ibid., no. 77 (1985).
5 Ibid., no. 80 (1985).
6 *Fish Farmer* 11, no. 1 (1988).
7 *AgScene*, no. 74 (1984).

8 Wordsworth, W. *Lines Composed above Tintern Abbey*.
9 Quoted by Thomas, K. *Man and the Natural World* (1983) p. 169.

Chapter 4 The Choice of Diet

1 Hewitt, J. A. 'Diet', in *Chambers's Encyclopaedia* (Chambers, London, 1950).
2 *American Heart Association Journal*, May 1984, quoted by Fox, M. *Agricide* (Schocken Books, New York, 1986).
3 COMA *Diet and Cardiovascular Disease* (HMSO, London, 1984).
4 *Statistical Abstract of the USA* (US Bureau of the Census, Washington DC, 1988).
5 *MAFF Statistics* (Government Statistical Service).
6 Fox, M. *Agricide* (1986) pp. 103–5.
7 London Food Commission. *Food Facts* (London Food Commission, London, 1987).
8 Chickens' Lib. Fact sheets, no. 387, 388.
9 Crawford, M. A. 'Re-evaluation of the Nutrient Role of Animal Products', in Reid, R. L. (ed.) III World Conference on Animal Production (Sydney University Press, Sydney, 1975) pp. 21–35.
10 London Food Commission. *Food Adulteration* (Unwin Hyman, London, 1988) p. 39.
11 Ibid., p. 43.
12 Ibid., p. 46.
13 Ibid., p. 54.
14 Clough, A. H. 'The Latest Decalogue', *The Poems of Arthur Hugh Clough* (Clarendon Press, Oxford, 1974) p. 205.
15 London Food Commission. *Food Adulteration* p. 175.
16 Ibid., p. 162.
17 Gerrard, F. *Meat Technology* (London, 1977).
18 Bernarde, M. A. *The Chemicals We Eat* (McGraw-Hill, New York, 1975) p. 108.
19 Forman, D., Al-Dabbagh, S. and Doll, R. 'Nitrates, Nitrites and Gastric Cancer in Great Britain' *Nature*, 313 (1985) pp. 620–5.
20 Drummond, J. C. and Wilbraham, A. 'Food, History of', in *Chambers's Encyclopaedia* (Chambers, London, 1950).
21 Hall, R. L. 'Food Additives', in Clydesdale, F. M. (ed.) *Food, Nutrition and You* (Prentice-Hall, Englewood Cliffs, NJ, 1977).
22 Abraham, J. and Millstone, E. 'Food Additive Controls: some international comparisons'. *Food Policy* 14, no. 1 (1989) p. 43.

Chapter 5 Free Range or Factory?

1 *Poultry World*, 28 March 1985.
2 Petch, A. Personal communication.

3 Crawford, M. A. 'Meat as a Source of Lipids', in Cole, D. J. A. and Lawrie, R. A. (eds) *Meat* (London, 1975).
4 Tolan, A. *British Journal of Nutrition* 31 (1974) p. 185.
5 Fox, M. *Agricide* (Schocken Books, New York, 1986) p. 35.
6 *AgScene*, no. 97 (1989).
7 Fox, M. *Agricide* (1986) p. 132.
8 Food Advisory Committee. *Final Report on the Review of the Colouring Matter in Food Regulations 1973* (1987).
9 Torrissen, D. J. 'Pigmentation of Salmonids'. *Aquaculture* 53 (1986) p. 271.
10 Walsh, J. *The Meat Machine* (Columbus, London, 1986).
11 Gerrard, F. *Meat Technology* (London, 1977) p. 115.
12 Hanrahan, J. P. (ed.) *Beta-agonists and their Effects on Animal Growth and Carcass Quality* (Elsevier Applied Science, London, 1987).
13 Flint, D. J. *Journal of Endocrinology* 115 (1987) p. 365.
14 *AgScene*, no. 94 (1989).
15 Flint, D. J., Futter, C. E. and Peaker, M. *News in Physiological Science* 2 (1987) p. 1.
16 James, W. *The Lancet* I (1987) p. 216.
17 *Meat Trades Journal*, 8 September 1988.
18 Lloyd, M. M. et al. *The Lancet* I (1987) p. 561.
19 Rodriguez, C. 'Environmental hormone contamination in Puerto Rico'. *New England Journal of Medicine* 310 (1984) p. 1741.
20 Fox, M. *Agricide* p. 79.
21 Muirhead, M. R. 'Porcine Pneumonia'. *Pig Veterinary Society Proceedings* 6 (1980) p. 103.
22 *Veterinary Record* 117 (1985) p. 345.
23 Agricultural and Food Research Council. *Annual Report of the Institute for Animal Health* (AFRC, London 1987) p. 27.
24 Threlfall, R. J. et al. 'Increasing incidence of resistance to gentamicin and related aminoglycerides in *Salmonella typhimurium* phage type 204c'. *Veterinary Record* 17 (1985) pp. 355–7.
25 The *Guardian*, 1 December 1989.
26 Fox, M. *Agricide*, p. 11.
27 Ibid., p. 71.
28 National Research Council. *Regulating Pesticides in Food* (National Academy Press, Washington, 1987).
29 Fox, M. *Agricide*, p. 67.
30 Ludvigsen, J. B. et al. *Livestock Production Science* 9 (1982) p. 65.
31 MAFF. *Food Surveillance Paper* no. 16 (HMSO, London, 1986).
32 *Farmers Weekly*, 19 May 1989.
33 Hayes, J. R. and Borzella, J. F. 'Biodeposition of environmental chemicals by animals', in Beitz, D. C. and Hansen, R. G. (eds) *Animal Products in Human Nutrition* (New York, 1982).

Chapter 6 In Sickness and In Health

1 Hobbs, B. C. and Gilbert, R. J. *Food Poisoning and Food Hygiene* (E. J. Arnold, London, 1978) p. 51.
2 Ryan et al. *Journal of the American Medical Association* 258 (1987) p. 3269.
3 London Food Commission. *Food Adulteration* (Unwin Hyman, London, 1988) p. 257.
4 Roberts, D., Boag, K., Hall, L. M. and Shipp, C. R. 'The isolation of salmonellas from British pork sausages and sausage meat' *Journal of Hygiene* 75 (1975) p. 173.
5 *New Scientist*, 2 October 1986.
6 Hobbs, B. C. and Roberts, D. *Food Poisoning & Food Hygiene* (London, 1987).
7 Wathes, C. M. et al. 'Aerosol infection of calves and mice with *Salmonella typhimurium*'. *Veterinary Record* 118 (1986) p. 240.
8 London Food Commission. *Food Adulteration* (1988) p. 247.
9 'Infectious diseases'. *OPCS Monitor*, July–September 1988.
10 Ibid.
11 Agriculture Committee, '1st Report Salmonella in eggs'. House of Commons Papers (HMSO, London, 1988–9, 108-I, p. xiii.
12 Free Range Egg Association. Newsletter, August 1989.
13 MAFF. Letter of 16 November 1989.
14 Galbraith, N. S. et al. 'The changing pattern of foodborne disease in England and Wales'. *Public Health* 101 (1987) pp. 319–28.
15 Gilbert, R. J. 'Foodborne infections and intoxication: recent problems and new organisms', in *CFPRA Symposium on Microbiological & Environmental Health Problems* 19–21 January 1987.
16 Neill, S. D., McLoughlin, M. and McIlroy, S. 'Type C botulism in cattle being fed ensiled poultry litter'. *Veterinary Record* 124 (1989) p. 558.
17 Roberts, T. A. 'Healthy animals and meat hygiene'. Address to Royal Society of Health, *Safety and Nutritional Value of Red Meat* (London, 24 November 1987).
18 Schlech, W. F. III, et al. 'Epidemic listeriosis – evidence for transmission by food'. *New England Journal of Medicine* 308 (1983) pp. 203–6.
19 El-Kest, S. E. and Marth, E. H. '*Listeria monocytogenes* and its inactivation by chlorine'. *Lebensmittel Wissenschaft und Technologie* 21 (1988) pp. 346–52.
20 Ibid.
21 Low. J. C. and Renton, C. P. 'Septicaemia, encephalitis and abortion in a housed flock of sheep caused by *listeria moncytogenes* type 1/2'. *Veterinary Record* 116 (1985) pp. 147–50.
22 Fenlon, D. R. 'Rapid quantitative assessment of the distribution of listeria in silage'. *Veterinary Record* 118 (1986) pp. 240–2.
23 *The Lancet*, 14 January 1989, pp. 83–4; 11 February 1989.

24 *Communicable Diseases Scotland*, 15 July 1989.
25 Gracey, J. F. *Meat Hygiene* 8th edition (Baillière, Tindall, London, 1986) p. 244.
26 Gold, M. *Assault and Battery* (Pluto Press, London, 1983) p. 23.
27 Donham, K. J., Berg, J. W. and Sawin, R. S. 'Epidemiological relationships of the bovine population and human leukemia in Iowa'. *American Journal of Epidemiology* 112 (1980) pp. 80–92.
28 Fox, M. *Agricide* (Schocken Books, New York, 1986) p. 95.
29 The *Observer*, 11 December 1988.
30 The *Independent*, 10 October 1988.
31 Holt, T. A. and Phillips, J. 'Bovine spongiform encephalopathy'. *British Medical Journal* 296 (1988) pp. 1581–2.
32 The *Independent*, 14 June 1989.
33 *Farmers Weekly*, 13 October 1989.
34 The *Guardian*, 10 November 1989.
35 Scholtissek, C. and Naylor, E. 'Fishfarming and influenza pandemics'. *Nature* 331 (1988) p. 215.
36 Shortridge, K. F. and Stuart-Harris, C. H. *The Lancet* II (1982) p. 812.
37 Scholtissek, C. and Naylor, E. 'Fishfarming' (1988).
38 Devick, O. (ed.). *Harvesting Polluted Waters* (New York, 1976) p. 205.
39 'Freshwater aquaculture development in China'. *Fisheries Technical Paper* no. 215 (FAO, Rome, 1983).
40 Swann Report. *Use of Antibiotics in Animal Husbandry and Veterinary Medicine* (HMSO, Cmnd 4190, 1969/70).
41 Threlfall, E. J. and Rowe, B. 'Antimicrobial drug resistance in salmonellae in Britain', in Woodbine, M. (ed.) *Antimicrobials and Agriculture* (Butterworths, London, 1984) p. 513.
42 Helmuth, R. et al. 'Gentamicin resistant salmonellae in turkey rearing', in Woodbine, M. *Antimicrobials* (1984) pp. 237–42.
43 Threlfall, E. J. et al. 'Increasing incidence of resistance to gentamicin and related aminoglycerides in *Salmonella typhimurium* phage type 204c'. *Veterinary Record* 17 (1985) pp. 355–7.
44 Threlfall, E. J. et al. 'Multiple drug-resistant strains of *Salmonella typhimurium* in poultry'. *Veterinary Record* 124 (1989) p. 538.
45 Threlfall, E. J. and Rowe, B. in Woodbine, M. *Antimicrobials* (1984).
46 Holmberg, S. and O'Brien, T. 'Drug resistant salmonella from animals fed antimicrobials'. *New England Journal of Medicine* 311 no. 10 (1984) pp. 617–22.
47 Lacey, R. W. 'Are resistant bacteria from animals and poultry an important threat to the treatment of human infection?' in Jolly, D. W. (ed.) *Ten Years on from Swann* (Association of Veterinarians in Industry, London, 1981) pp. 127–44.
48 Wierup, M. 'Human and animal consumption of antibiotics and chemotherapeutic drugs', in Woodbine, M. *Antimicrobials* (1984) pp. 483–9.
49 Guinee, P. et al. '*E. coli* with resistance factors in vegetarians, babies and non-vegetarians'. *Applied Microbiology* 50 (1970) pp. 531–5.

50 Linton, K. B., Lee, P. A., Richmond, M. H., Gillespie, W. A., Rowland, A. J. and Baker, V. N. 'Antibiotic resistance and transmissible R-factors in the intestinal coliform flora of healthy adults and children in an urban and a rural community'. *Journal of Hygiene* 70 (1972) pp. 99–104.
51 London Food Commission. *Food Adulteration* (1988) pp. 190–233.
52 Priyadarshini, I. and Tulpule, P. 'Effects of graded doses of gamma irradiation on aflatoxin production by Aspergillus parasiticus in wheat'. *Food and Cosmetic Toxicology* 17 (1979) p. 505.
53 Josephson, E. S. and Peterson, M. S. (eds). *Preservation of Food by Ionizing Radiation* (CRC Press, Florida, 1982).
54 USDA Food Safety and Inspection Service. *Scheme and Critical Variables for a Limited Study on the Effects of Vacuum Packaging on Irradiated Pork Loins* (GPO, Washington DC, 1986).
55 London Food Commission. *Food Adulteration* (1988) p. 223.
56 The *Guardian*, 6 October 1989.
57 Humphrey, N., in Ferry, G. (ed.) *The Understanding of Animals* (Basil Blackwell, Oxford, 1984) p. 243.
58 The *Guardian*, 11 November 1986.

Chapter 7 Why Care About Animals?

1 Suetonius. *The Twelve Caesars* (Penguin, Harmondsworth) p. 297.
2 Bentham, J. *Introduction to the Principles of Morals and Legislation* (Athlone Press, London, 1970) p. 283.
3 Singer, P. *Animal Liberation* (London, 1976).
4 Regan, T. *The Case for Animal Rights* (RKP, London, 1983).
5 Midgley, M. *Animals and Why They Matter* (Penguin, Harmondsworth, 1983).
6 Clark, S. R. L. *The Moral Status of Animals* (OUP, Oxford, 1977).
7 Kellert, S. R. and Westervalt, M. O. 'Historical trends in American animal use and perception'. *Transactions of the 47th North American Wildlife & Natural Resources Conference* (1982).

Chapter 8 The Assessment of Animal Welfare

1 Descartes, R. *Discourse on Method* (Dent, London) p. 27.
2 Eyewitness account of experiments at Port Royal, quoted by Singer P. *Animal Liberation* (London, 1976) p. 220.
3 RSPCA. *Pain and Suffering in Experimental Animals in the United Kingdom* (RSPCA, Horsham, 1983) p. 6.
4 RSPCA. *Report of the Panel of Enquiry into Shooting and Angling* (RSPCA, Horsham, 1980).
5 Shakespeare. *Measure for Measure*, Act 3, sc. 1.

6 Wittgenstein, L. *Philosophical Investigations* (Blackwell, Oxford, 1968) §281ff.
7 Sanford, J., Ewbank, R., Molony, V., Tavernor, W. D. and Uvarov, D. 'Guidelines for the Recognition and Assessment of Pain in Animals'. *Veterinary Record* 118 (1986) pp. 334–8.
8 *New Scientist*, 2 April 1987.
9 Lorenz, K. *Behind the Mirror* (Methuen, London, 1977) p. 55ff.
10 Brambell, F. W. R. *Report of the Technical Committee to Enquire into the Welfare of Animals kept under Intensive Livestock Husbandry Systems.* Cmnd 2836 (HMSO, London, 1965).
11 Wood-Gush, D. G. M. *Elements of Ethology* (Chapman and Hall, London, 1983).
12 Kiley-Worthington, M. *The Behaviour of Horses* (J. A. Allen, London, 1987).
13 Durrell, G. *The Stationary Ark* (Collins, London, 1976).
14 Morris, D. 'The response of animals to a restricted environment'. *Symposium of the Zoological Society of London* 13 (1964) p. 99.
15 Hediger, H. *Wild Animals in Captivity – An Outline of the Biology of Zoological Gardens* (Dover, New York, 1964).
16 Broom, D. M. 'The scientific assessment of animal welfare'. *Applied Animal Behaviour Science* 20 (1988) pp. 5–19.
17 Andreae, U. and Smidt, D. 'Behavioural alterations in young cattle on slatted floors'. *Hohenheimer Arbeiten* 121 (1982) pp. 51–60.
18 Selye, H. *The Physiology and Pathology of Exposure to Stress* (Montreal, 1950).
19 Baldock, N. M. and Sibly, R. M. 'Effects of management procedures on heart rate in sheep'. *Applied Animal Behaviour Science* (1988).
20 Duncan, I. J. H. *Applied Animal Behaviour Science* 16 (1986) p. 97.
21 Broom, D. M., Knight, P. G. and Stansfield, S. C. 'Hen behaviour and hypothalamic-pituitary-adrenal responses to handling and transport'. *Applied Animal Behaviour Science* 16 (1986) p. 98.
22 Dawkins, M. S. *Animal Suffering* (London, 1980) p. 49.
23 Broom, D. M. 'Scientific assessment of animal welfare' (1988).
24 Bowsher, D. 'Pain sensations and pain reactions', in Wood-Gush, D. G. M. et al. *Self-awareness in Domesticated Animals* (UFAW, Potters Bar, 1981) p. 25.
25 Cronin, G. M. and Wiepkema, P. R. 'An analysis of stereotyped behaviour in tethered sows'. *Annales de Recherches Vétérinaires* 15 (1984) pp. 263–70.
26 UFAW. Report and Accounts 1987–88.
27 Dawkins, M. S. *Animal Suffering* (1980) p. 89.
28 Duncan, I. J. H. 'Animal behaviour and welfare', in Clark, J. A. (ed.) *Environmental Aspects of Housing for Animal Production* (Butterworths, London, 1981) p. 461.
29 M. Kiley-Worthington, personal communication.

Chapter 9 Welfare in Modern Farming

1 Noren, O. 'Noxious gases and odours', in Taiganides, E. P. (ed.). *Animal Wastes* (London, 1977) p. 114.

2 Dawkins, M. S. 'Time budgets in Red Jungle fowl as a baseline for the assessment of welfare in domestic fowl'. *Applied Animal Behaviour Science* 24 (1989) pp. 77–80.

3 *Poultry World*, 14 June 1984.

4 Ibid., March 1988.

5 Hughes, B. O. and Black, A. J. 'The preference of domestic hens for different types of battery cage floor'. *British Poultry Science* 14 (1973) p. 615.

6 Dawkins, M. S. 'Towards an objective method of assessing welfare in domestic fowl'. *Applied Animal Ethology* 2 (1976) p. 245.

7 McLean K. A., Baxter, M. R. and Michie, W. 'A comparison of the welfare of laying hens in battery cages and in a perchery'. *R & D in Agriculture* 3 no. 2 (1986) pp. 93–8.

8 Dawkins, M. S. 'Space needs of laying hens'. *RSPCA Today*, Summer 1986 p. 21.

9 Hughes, B. O. *British Poultry Science* 18 (1977) pp. 9–18.

10 Duncan, I. J. H. 'Animal behaviour and welfare', in Clark, J. A. (ed.) *Environmental Aspects of Housing for Animal Production* (Butterworths, London, 1981) pp. 463–5.

11 Squires, E. J. and Leeson, S. 'Aetiology of fatty liver syndrome in laying hens'. *British Veterinary Journal* 144 no. 6 (1988) p. 602.

12 *Poultry Tribune*, February 1987 p. 40.

13 Quoted by Chickens' Lib in publicity leaflet.

14 Farm Animal Welfare Council. *Egg Production Systems – An Assessment* (FAWC, Surbiton, 1986).

15 Van Putten, G. 'Ever been close to a nosey pig?'. *Applied Animal Ethology* 5 (1979) p. 298.

16 Scottish Farm Buildings Investigation Unit. *Does Close Confinement Cause Distress in Sows?* (Aberdeen, 1986).

17 *Pig Farming*, January 1989.

18 *AgScene*, no. 88 (1987).

19 Hunt, K. and Petchey, T. 'The behaviour and environmental preferences of sows around farrowing'. *RSPCA Today*, Summer 1987.

20 *AgScene*, no. 79 (1985).

21 Farm Animal Welfare Council. *Assessment of Pig Production Systems* (FAWC, Surbiton, 1988).

22 *The Vealer*, April 1984.

23 Kiley-Worthington, M. *Behavioural Problems of Farm Animals* (Oriel Press, Stocksfield, 1977) p. 33.

24 Andreae, U. and Smidt, D. 'Behavioural alterations in young cattle on slatted floors'. (1982) pp. 51–60.

25 Hannan, J. A. and Murphy, P. A. 'Comparative mortality and morbidity rates for cattle on slatted floors and in straw yards', in Smidt, D. (ed.). *Indicators Relevant to Farm Animal Welfare* (Nijhoff, The Hague, 1983).
26 Webster, J. *Understanding the Dairy Cow* (BSP Professional, Oxford, 1987) p. 99.
27 Kiley-Worthington, M. *Behavioural Problems* (1977) p. 31.
28 Ibid. pp. 30, 35.
29 Wood-Gush, D. G. M. and Marsden, D. 'A pen is not just a pen to a sheep'. *RSPCA Today*, Spring 1986.
30 Farm Animal Welfare Council. *Report on Priorities in Animal Welfare Research and Development* (FAWC, Surbiton, 1988).
31 *New Scientist*, 2 April 1987.
32 *Animals International*, Winter 1988.
33 Squires, E. J. and Leeson, S. 'Aetiology of fatty liver syndrome' (1988).
34 Gentle, M. J. and Hill, F. L. 'Oral lesions in the chicken: behavioural responses following nociceptive stimulation'. *Physiology and Behaviour* 40 (1987) pp. 781–3.
35 Timms, L. M. 'Forms of leg abnormality in male broilers fed on a diet containing 12.5% rapeseed meal' *Research in Veterinary Science* 35 (1983) pp. 182–9.
36 *AgScene*, no. 74 (1984).
37 *Poultry World*, May 1987.
38 Agricultural Committee. 'Animal welfare in poultry, pig and veal calf production' (minutes of evidence) *House of Commons Papers* 38–i (1980–81) p. 106.
39 *AgScene*, no. 74 (1984).
40 Ibid., no. 83 (1986).
41 The *Independent*, 15 May 1989.
42 Farm Animal Welfare Council. *Report on Priorities* (FAWC, Surbiton, 1988).
43 Berg, R. T. and Butterfield, R. M. 'New and improved types of meat animal', in Lawrie, R. (ed.), *Developments in Meat Science* – 3 (Elsevier Applied Science, London, 1985) p. 11.
44 *Poultry World*, October, 1986.
45 MAFF. *Codes of Recommendations for the Welfare of Livestock: Domestic Fowls*, leaflet 703 (1987); *Turkeys*, leaflet 704 (1987).
46 Gentle, M. J., Waddington, D., Hunter, L. N. and Jones, R. B. 'Behavioural evidence for persistent pain following partial beak amputation in chickens'. *Applied Animal Behaviour Science* 27 (1990) pp. 149–57.
47 Soil Association. *Standards for Organic Agriculture* (Bristol, 1987).
48 The Welfare of Livestock (Prohibited Operations) Regulations 1982 Statutory Instrument no. 1884 (HMSO, London, 1982).
49 Ibid.
50 *AgScene*, no. 84 (1986).
51 RSPCA. *The Slaughter of Food Animals* (RSPCA, Horsham, 1986).
52 Grandin, T. *Int. J. Stud. Anim. Prob.* 1 (1980) pp. 121, 178.

53 Everton, A. R. *Veterinary Record* 110 (1982) p. 469.
54 Farm Animal Welfare Council. *Report on the Welfare of Poultry at the Time of Slaughter* (FAWC, Surbiton, 1982); *Report on the Welfare of Livestock (Red Meat Animals) at the Time of Slaughter* (FAWC, Surbiton, 1984).
55 Leach, T. M. 'Pre-slaughter stunning', in Lawrie, R. *Developments in Meat Science* (1985) p. 66.
56 Newhook, J. C. and Blackmore, D. K. *Meat Science* 6 (1982) p. 295.
57 For further discussion see the UFAW Symposium *Humane Slaughter of Animals for Food* (UFAW, South Mimms, 1987).
58 Masri, Al-Hafiz, B. A. *Islamic Concern for Animals* (The Athene Trust, Petersfield, 1987).
59 *AgScene*, no. 77 (1985).
60 Ibid., no. 91 (1988).
61 British Veterinary Association Animal Welfare Foundation. *Symposium on the Welfare of Animals in Transit* (1987).
62 EC Directive on the Protection of Animals During International Transport (EC 77–489).
63 The Welfare of Poultry (Transport) Order 1988.
64 The Markets (Protection of Animals) Order 1986.
65 Farm Animal Welfare Council. *Welfare of Livestock at Markets* (FAWC, Surbiton, 1986).
66 *AgScene*, no. 84 (1986).
67 McDonald, C. L. 'Research into behaviour, nutrition and health of sheep during live export'. *Proceedings of the Australian Society of Animal Producers* 16.
68 *AgScene*, no. 84 (1986).
69 Ibid., no. 91 (1988).
70 *RSPCA Today*, Winter 1988.
71 Ibid., Spring 1988.
72 Webster, J. *Dairy Cow* (1989) p. 109.
73 Kiley-Worthington, M. *Behavioural Problems* p. 96.

Chapter 10 Livestock in the Countryside

1 Countryside Commission. *New Agricultural Landscapes* (Countryside Commission, London, 1977).
2 Turner, J. *The Politics of Landscape: Rural scenery and society in English poetry 1630–1660* (Basil Blackwell, Oxford, 1979) p. 165.
3 *AgScene*, no. 89 (1987).
4 Sainsbury, D. W. B. 'Health problems in intensive animal production', in Clark, J. A. (ed.), *Environmental Aspects of Housing for Animal Production* (Butterworths, London, 1981) p. 441.
5 *Royal Commission on Environmental Pollution Seventh Report: Agriculture and Pollution* Cmnd 7644 (HMSO, London, 1979) §5.21.

6 *AgScene*, no. 75 (1984).
7 Fox, M. *Agricide* (Schocken Books, New York, 1986) p. 61.
8 Van Burg, P. J. F. et al. 'Nitrogen and intensification of livestock farming in EEC countries'. *Proceedings of the Fertiliser Society* no. 199, London, 23 April 1981.
9 Water Authorities Association. *Water Pollution from Farm Waste 1987, England and Wales* (Water Authorities Association, 1988) p. 38.
10 *Royal Commission on Environmental Pollution Seventh Report* §5.75.
11 Water Authorities Association. *Water Pollution 1987* (1988) p. 12.
12 *Royal Commission on Environmental Pollution Seventh Report* §5.14.
13 Water Authorities Association. *Water Pollution 1987* (1988) p. 3.
14 *Royal Commission on Environmental Pollution Seventh Report* §5.16.
15 Strauch, D. 'Management of hygienic problems in large animal feedlots', in Taiganides, E. P. (ed.), *Animal Wastes* (Applied Science Publishers, London, 1977) p. 95.
16 Ibid., p. 96.
17 *Royal Commission on Environmental Pollution Seventh Report* §5.20.
18 Ibid., §5.21.
19 Strauch, D. 'Management of hygienic problems' (1977) p. 101.
20 Ibid., p. 97.
21 Ibid., p. 99.
22 Kurc, R. 'Land disposal of feedlot wastes by irrigation in Czechoslovakia', in Taiganides, E. P. (ed.), *Animal Wastes* (1977) p. 329.
23 *Royal Commission on Environmental Pollution Seventh Report* Table 2.5.
24 Jones, P. H. 'Criteria and guidelines for the selection of animal feedlot sites', in Taiganides, E. P. (ed.), *Animal Wastes* (1977) p. 43.
25 Cajakada, E. 'Agriculture in Czechoslovakia'; Farkas, P. and Racz, T. 'Feedlot waste management in Hungary'; Runov, B. A. 'Feedlot waste management in the Soviet Union', in Taiganides, E. P. (ed.), *Animal Wastes* (1977) pp. 9, 392, 395.
26 *AgScene*, no. 89 (1987).
27 Kurc, R. 'Land disposal' (1977) p. 331.
28 *Farmers Weekly*, 6 October 1989.
29 Symes, D. and Marsden, T. 'Industrialisation of agriculture in intensive livestock farming in Humberside', in Healey, M. J. and Ilbery, B. W. (eds), *The Industrialisation of the Countryside* (Geo Books, Norwich, 1985).
30 Water Authorities Association. *Water Pollution 1987* (1988) p. 12.
31 *Royal Commission on Environmental Pollution Seventh Report* §5.22
32 Strauch, D. 'Management of hygienic problems' (1977) p. 102.
33 Taiganides E. P. 'Composting of feedlot wastes', in Taiganides, E. P. (ed.), *Animal Wastes* (1977) p. 241.
34 *Royal Commission on Environmental Pollution Seventh Report* §5.30.
35 Hojovec, J. 'Health effects from waste utilisation', in Taiganides, E. P. (ed.), *Animal Wastes* (1977) p. 106.
36 Ibid.
37 Ibid.

38 Jewell, W. J. and Loehr, R. C. 'Energy recovery from animal wastes', in Taiganides, E. P. (ed.), *Animal Wastes* (1977) p. 273.
39 Martin, J. H. 'Biogas production from manures: a realistic assessment', in Lockeretz, W. (ed.), *Agriculture as a Producer and Consumer of Energy* (Boulder, Colorado, 1982).

Chapter 11 Fish Farming and the Environment

1 Nature Conservancy Council. *Fishfarming and the Safeguard of the Natural Marine Environment of Scotland* (NCC, Edinburgh, 1989) p. 14.
2 Water Authorities Association. *Water Pollution from Farm Waste 1987, England and Wales* (Water Authorities Association, London, 1988) p. 32.
3 Solbé, J. F. de L. G. and Seager, J. 'Effluent Control', *Fish Farmer* May/June 1988, p. 11.
4 Ibid.
5 Niemi, M. and Taipalinen, I. 'Faecal indicator bacteria at fish farms'. *Hydrobiologia* 86 (1982) pp. 171–5.
6 Nature Conservancy Council. (1989) p. 41.
7 Solbé, J. F. de L. G. and Seager, J. 'Effluent Control' (1988) p. 11.
8 Lovelock, J. *Gaia: A New Look at Life on Earth* (OUP, Oxford, 1979).
9 Nishimura, A. *Bulletin of the Plankton Society of Japan* 29 (1982) pp. 1–7.
10 Nature Conservancy Council. *Fishfarming* (1989) p. 36.
11 Ibid., p. 37.
12 Assuming BOD ≃ 250 mg/kg fish/hr (NCC *Fishfarming* 1989, p. 37), 200 tonnes will require 50 kg oxygen/hr. If BOD of domestic sewage ≃ 1 g/l, this oxygen requirement would be supplied by 50,000 litres or about 10,000 gallons.
13 Nature Conservancy Council. *Fishfarming* (1989) p. 49.
14 *Fish Farmer* 11, no. 4 (1988) p. 15.
15 Dolmen, D. '*Gyrodactylus salaris* (Monogenea) in Norway; infestations and management', in Stenmark, R. and Malmberg, G. (eds), *Proceedings of the Symposium on Parasites and Diseases in Natural Waters and Aquaculture in Nordic Countries* (Stockholm, 1987) pp. 63–9.
16 Maitland, P. S. 'The impact of farmed salmon on the genetics of wild stocks'. Report to the Nature Conservancy Council, 1987 (unpublished).
17 Scottish Wildlife and Countryside Link. *Marine Fishfarming in Scotland – A Discussion Paper* (Scottish Wildlife and Countryside Link, Perth, 1988) p. 30.
18 Ibid.; also A. Mackay, personal communication.
19 *Fish Farmer* 11, no. 6 (1988).
20 Austin, B. 'Chemotherapy of bacterial fish diseases', in Ellis, A. E. (ed.), *Fish and Shellfish Pathology* (Academic Press, London, 1985) pp. 19–26.
21 Nature Conservancy Council. *Fishfarming* (1989) p. 102.
22 *Fish Farmer* 11, no. 5 (1988).

23 Egidius, E. and Møster, B. 'Effect of neguvon and nuvan treatment on crabs, lobsters and blue mussel'. *Aquaculture* 60 (1987) pp. 165–8.
24 The *Observer*, 1 January 1989; *West Highland Free Press*, 30 June 1989.
25 *West Highland Free Press*, 8 January 1988.
26 Nature Conservancy Council. *Fishfarming* (1989) p. 101.
27 Scottish Wildlife and Countryside Link. *Marine Fishfarming* (1988) p. 21.
28 Ross, A. *Controlling Nature's Predators on Fish Farms* (Marine Conservation Society, Ross-on-Wye, 1988).
29 Carss, D. N. *The Effects of Piscivorous Birds on Fish Farms on the West Coast of Scotland* (PhD thesis, University of Edinburgh, 1988).
30 Nature Conservancy Council. *Fishfarming* (1989) p. 82; p. 73.
31 Ross, A. *Controlling Nature's Predators* (1988).
32 *Aberdeen Press and Journal*, 18 March 1988.
33 Nature Conservancy Council. *Fishfarming* (1989); Ross, A. *Controlling Nature's Predators* (1988).
34 Nature Conservancy Council. *Fishfarming* (1989) pp. 75–81.
35 Ibid., pp. 77, 85.
36 Ibid., p. 72.
37 *Fish Farmer* 11, no. 6 (1988) p. 46.

Chapter 12 Farming for a Healthy Planet

1 Thomas, K. *Man and the Natural World: Changing Attitudes in England 1500–1800* (Allen Lane, London, 1983) pp. 19, 20.
2 International Union for the Conservation of Nature and Natural Resources. *World Conservation Strategy, Living Resource Conservation for Sustainable Development* (IUCN-UNEP-WWF, 1980).
3 Colinvaux, P. *Why Big Fierce Animals are Rare* (Penguin, London, 1980).
4 Lappe, F. Moore *Diet for a Small Planet* (New York, 1971) p. 10.
5 *AgScene*, no. 77 (1985).
6 Annual report on Agriculture – FY1987 (Foreign Press Center, Japan, 1988).
7 *AgScene*, no. 77 (1985).
8 Ibid.
9 Nature Conservancy Council. *Fishfarming and the Safeguard of the Natural Marine Environment of Scotland* (NCC, Edinburgh, 1989) p. 93.
10 Heubeck, M. (ed.), *Proceedings of a Seminar on Seabirds and Sandeels* (Shetland Bird Club, Lerwick, January 1988).
11 Data from *Fish Farmer* and NCC *Fishfarming* (1989).
12 Fox, M. *Farm Animal Welfare and the Human Diet* (Humane Society of the United States, Washington, 1988) p. 26.
13 Teutscher, F. 'Fish, food and human nutrition'. *Food and Nutrition* 12, no. 12 (1986) p. 2.
14 Amery, J. *To Die So Young*, quoted in Chickens' Lib fact sheets.
15 *AgScene*, no. 90 (1987).

16 Webster, J. *Understanding the Dairy Cow* (BSP Professional, Oxford, 1989) p. 15.
17 Borgstrom, G. *The Hungry Planet* (London, 1965).
18 Tudge, C. *Food Crops for the Future* (Basil Blackwell, Oxford, 1988) p. 43.
19 US Dept of Agriculture. *Economics of Scale in Farming* (GPO, Washington DC, 1967).
20 Bowler, I. 'Some consequences of the industrialisation of agriculture in the EEC', in Healey, M. J. and Ilbery, B. W. (eds), *The Industrialisation of the Countryside* (Geo Books, Norwich, 1985).
21 To regard other species merely as 'renewable resources' is, argues the leading Animal Rights philosopher Tom Regan, no better than viewing Jews as such a resource for powerful genitles, Blacks for avaricious Whites, or women for chauvinistic men. In the present context, however, economics and compassion lead equally to the conclusion that the extinction of species should be avoided.
22 Fox, M. *Agricide* (Schocken Books, New York, 1986) p. 51.
23 Ibid., p. 54.
24 Hartmans, J. L. 'Animal health in relation to intensive pasture use', in Reid, R. L. (ed.), III World Conference on Animal Production (Sydney University Press, Sydney, 1975) pp. 233–7.
25 Buffon. *Histoire Naturelle* translated by M. Cranston in Rousseau, J. J. *A Discourse on Inequality* (Penguin, Harmondsworth, 1984).
26 Fox, M. *Agricide* (1986) p. 54.
27 Bigalki, R. C. 'Utilisation of terrestrial wild animals', in Reid, R. L. (ed.), Conference on Animal Production (1975) pp. 36–46.
28 *Greenpeace News*, Autumn 1989.
29 Molion, L. C. B. 'The Amazonian forests and climatic stability'. *The Ecologist* 19, no. 6 (1989) pp. 211–13.
30 Fox, M. *Agricide* (1986) p. 61.
31 Farm and Food Society. *Agriculture and Pollution* (Farm and Food Society, London, 1980).

Chapter 13 Politics and Profits

1 *Hansard*, 1 November 1983 pp. 341–2.
2 Cottrell, R. *The Sacred Cow* (Grafton, London, 1987) p. 126.
3 Body, R. *Farming in the Clouds* (Temple Smith, London, 1984) p. 43.
4 Ibid., p. 31.
5 Buckwell, A. 'Do we benefit from CAP? – A review', in Ross, D. B. and Sellers, K. C. (eds), *The Effects of EEC CAP on Animal Production and Veterinary Medicine.* Symposium, Association of Veterinarians in Industry, London, 7/8 October 1985 (Association of Veterinarians in Industry, London, 1986).
6 Body, R. *Farming* (1984) p. 38.
7 *Farmers Weekly*, 6 October 1989.

8 Buckwell, A. 'Do we benefit from CAP?' (1985) p. 39.
9 *Farmers Weekly* 20 January 1984.
10 Body, R. *Farming* (1984) pp. 54–6.
11 Ibid., p. 53.
12 Schumacher, E. F. *Small is Beautiful* (London, 1973).
13 Sainsbury, D. and Sainsbury, P. *Livestock Health and Housing* (London, 1979) p. 2.
14 Kiley-Worthington, M. *Behavioural Problems of Farm Animals* (Oriel Press, Stocksfield, 1977) pp. 36–7.
15 Ibid., p. 104.
16 Carnell, P. *Alternatives to Factory Farming* (Earth Resources, London, 1983).
17 Hill, J. A. *R & D in Agriculture* 3, no. 1 (1986) p. 13.
18 Carnell, P. *Alternatives* (1983).
19 *MLC Commercial Pig Production Yearbook 1979*, cited by Carnell, P. *Alternatives* (1983) p. 26.
20 *National Agricultural Centre Profile* 1987/8 (National Agricultural Centre, Stoneleigh, Warwickshire) p. 30.
21 Carnell, P. *Alternatives* (1983) pp. 27–8.
22 Webster, J. *Understanding the Dairy Cow* (BSP Professional, Oxford, 1989) p. 332.
23 Donham, K. J., Zavala, D. C. and Merchant, J. A. 'Respiratory symptoms and lung function among workers in swine confinement buildings'. *Archives of Environmental Health* 39, no. 2 (1984) pp. 96–100.
24 Clark, S., Rylander, R. and Larrson, L. 'Airborne bacteria, endotoxin and fungi dust in poultry and swine confinement buildings'. *Amer. Ind. Hyg. Assoc.* 44, no. 7 (1983) pp. 537–41.

Chapter 14 Serving the Consumer?

1 Cannon, G. *The Politics of Food* (London, 1987) p. 61.
2 Ibid.
3 Ibid., p. 29.
4 Ibid., p. 46.
5 Orr, J. Boyd. *Food, Health and Income* (London, 1936).
6 Cannon, G. *Politics of Food* (1987) p. 50.
7 Ibid., pp. 254–6.
8 Webb, T. and Lang, T. *Food Irradiation – The Facts* (Wellingborough, 1987).
9 Abraham, J. and Millstone, E. 'Food additive controls: some international comparisons. *Food Policy* 14, no. 1 (1989) p. 43.
10 London Food Commission. *Food Adulteration* (Unwin Hyman, London, 1988) p. 24.
11 Arendt, H. 'Crises of the Republic', cited by Cannon, G. *Politics of Food* (1987) p. 221.

12 Cannon, G. *Politics of Food* (1987) pp. 65–71.
13 The *Independent*, 20 December 1988.
14 Body, R. *Farming in the Clouds* (Temple Smith, London, 1984) p. 152.
15 *Royal Commission on Environmental Pollution Seventh Report: Agriculture and Pollution Cmnd 7644 (HMSO, London, 1979)* §5.89.
16 Symes, D. and Marsden, T. 'Industrialisation of agriculture in intensive livestock farming in Humberside', in Healey, M. J. and Ilbery, B. W. (eds), *The Industrialisation of the Countryside* (Geo Books, Norwich, 1985).
17 Runov, B. A. 'Feedlot waste management in the Soviet Union', in Taiganides, E. P. (ed.), *Animal Wastes* (Applied Science Publishers, London, 1977) p. 397.
18 *AgScene*, no. 93 (1988).
19 *AgScene*, no. 92 (1988); no. 94, 97 (1989).
20 *West Highland Free Press*, 28 October 1989.

Chapter 15 Battle Joined

1 Quoted by Ryder, R. *Animal Revolution* (Basil Blackwell, Oxford, 1989) p. 23.
2 Serpell, J. *In the Company of Animals* (Basil Blackwell, Oxford, 1986).
3 Turner, E. S. *All Heaven in a Rage* (London, 1964).
4 *AgScene*, no. 93 (1988); no. 94 (1989).
5 The *Sunday Times Magazine*, 12 November 1989.
6 Harrison, R. Farm Animal Welfare, What, if Any Progress? UFAW – Hume Memorial Lecture 1987.
7 *AgScene*, no. 93 (1988).
8 Thomas, K. *Man and the Natural World: Changing Attitudes in England 1500–1800* (Allen Lane, London, 1983) p. 299.
9 Harrison, R. Farm Animal Welfare (1987).
10 *AgScene*, no. 88 (1987).
11 Ibid., no. 77 (1985).
12 *Veterinary Record*, 4 October 1986.
13 Ryder, R. *Animal Revolution* (1989) p. 204.
14 *AgScene*, no. 74 (1984).
15 *The Times*, 5 August 1988.

Chapter 16 The International Scene

1 Ryder, R. *Animal Revolution* (Basil Blackwell, Oxford, 1989) pp. 167–77.
2 *Animals International* 8, no. 26 (1988).
3 Rojahan, A. 'Animal welfare legislation', in Smidt, D. (ed.) *Indicators Relevant to Farm Animal Welfare* (Nijhoff, The Hague, 1983) p. 11.
4 *AgScene*, no. 93 (1988); no. 94 (1989).

5 Eurogroup for Animal Welfare. *Animal Welfare Laws in Europe* (Eurogroup for Animal Welfare, Brussels, 1989).
6 *AgScene*, no. 91, 93 (1988).
7 *AgScene*, no. 77 (1985).
8 Dekker, A. J. *The Rise and Subsequent Development of the Dutch Deep-litter Egg* (Dutch Society for the Protection of Animals – Nederlandse Vereniging Tort, The Hague).
9 Council decision 78/923/EEC 19 June 1978 (OJ L.323 17 November 1978).
10 European Convention no. 87 (Strasbourg, 10 March 1976).
11 Council Resolution of 22 July 1980 (OJ C.196 1 August 1980).
12 Council Directive 86/113/EEC (OJ L.95 10 April 1986) p. 45.
13 Ryder, R. *Animal Revolution* (1989) p. 293.
14 Working Document no. A2-211/86.
15 *AgScene*, no. 88 (1987).
16 Report in Bunte, translated in *AgScene*, no. 88 (1987).
17 Ana Isabel Pinto, personal communication, 1989.
18 *AgScene*, no. 97 (1989).
19 Ibid.
20 Masri, Al-Hafiz, B. A. *Islamic Concern for Animals* (The Athene Trust, Petersfield, 1987).
21 Ryder, R. *Animal Revolution* (1989) p. 296.

Chapter 17 Little by Little

1 *AgScene*, no. 96 (1989).
2 Commission Regulation no. 1943/85 12 June 1985.
3 *AgScene* no. 94 (1989).
4 *Pig Farming*, January 1989.
5 *AgScene*, no. 94 (1989).
6 Dutch SPCA. *Alternatives to the Battery Cage System for Laying Hens* (Nederlandse Vereniging Tot, The Hague, 1987).
7 Farm Animal Welfare Council. *Egg Production Systems – An Assessment* (FAWC, Surbiton, 1986).
8 Webster, J. *Understanding the Dairy Cow* (BSP Professional, Oxford, 1989) p. 209ff.
9 *Pig Farming*, March 1989.
10 MAFF. *Codes of Recommendations for the Welfare of Livestock* (HMSO, London, various years).
11 Chickens' Lib. Fact sheets.
12 Gunn, S. D. 'The impact of CAP on veterinary practice', in Ross, D. B. and Sellers, K. C. (eds), *The Effects of EEC CAP on Animal Production and Veterinary Medicine*. Symposium, Association of Veterinarians in Industry, London, 7/8 October 1985.
13 *AgScene*, no. 92 (1988).

14 Wookey, B. *Rushall – The Story of an Organic Farm* (Basil Blackwell, Oxford, 1987).
15 Soil Association. *Standards for Organic Agriculture* (Bristol, 1987).

Appendices

1 Steering Group on Food Surveillance Papers. *No. 8 Survey of Arsenic in Food*; *No. 10 Survey of Lead in Food*; *No. 17 Survey of Mercury in Food* (HMSO, London).
2 Steering Group on Food Surveillance Paper. *No. 25 Report of the Working Party on Pesticide Residues 1985–8* (HMSO, London, 1989).
3 Ruiter, A. 'Contaminants in meat and meat products', in Lawrie, R. (ed.), *Developments in Meat Science – 3* (Elsevier Applied Science, London, 1985) pp. 185–220.
4 Ramade, F. *Ecotoxicology* (Wiley, Chichester, 1987).
5 Steering Group on Food Surveillance Paper. *No. 18. Mycotoxins* (HMSO, London, 1987).
6 Armstrong, D. G. 'Antibiotics as feed additives for ruminant livestock', in Woodbine, M. (ed.), *Antimicrobials and Agriculture* (Butterworths, London, 1984) pp. 331–47.
7 Dewdney, J. M. and Edwards, R. G. 'Penicillin hypersensitivity – Is milk a significant hazard?', in Woodbine, M. *Antimicrobials* (1984) pp. 464–5.
8 Ziv, G. 'Therapeutic use of antibiotics in farm animals', in Moats, W. A. (ed.), *Agricultural Uses of Antibiotics* (Washington, 1987) p. 18.
9 Steering Group on Food Surveillance Paper. *No. 22. Anabolic, Anthelmintic and Antimicrobial Agents* (HMSO, London, 1987).
10 Fox, M. *Agricide* (Schocken Books, New York, 1986) p. 81.
11 Food Advisory Committee. *Final Report on the Review of the Colouring Matter in Food Regulations 1973* (HMSO, London, 1987).
12 Lyne, A. R. and Lott, A. F. 'Inhibitory substances in animal feeds', in Woodbine, M. *Antimicrobials* (1984) pp. 413–22.
13 Walters, C. L. 'Nitrosamines in meat products', in Lawrie, R. (ed.), *Developments in Meat Science – 1* (London, 1980) p. 195.
14 London Food Commission. *Food Adulteration* (Unwin Hyman, London, 1988) p. 54.
15 Ibid., *passim*.
16 Cunningham, H. M. and Lawrence, G. A. 'Effect of exposure of meat and poultry to chlorinated water'. *Journal of Food Science* 42 (1977) p. 1504.
17 Steering Group on Food Surveillance Paper. *No. 21 Survey of Plasticiser Levels in Food Contact Materials and in Foods* (HMSO, London, 1987).
18 Elias, P. S. 'Irradiation preservation of meat and meat products', in Lawrie, R. *Meat Science* (1985) pp. 115–54.
19 *OPCS. Monitor* (1988 figures).
20 Threlfall, E. J. and Rowe, B. 'Antimicrobial drug resistance in salmonellae

in Britain', in Woodbine, M. *Antimicrobials* (1984) p. 515 estimate approximately 10 per cent of human salmonella infections are multi-resistant. The prevalent strains change so quickly, that statistics are inevitably very approximate.

21 Extrapolated from *Communicable Diseases Scotland* (1989).

22 Estimated from Hobbs, B. C. and Roberts, D. *Food Poisoning & Food Hygiene* (London, 1987) p. 34.

23 Nature Conservancy Council. *Fishfarming and the Safeguard of the Natural Marine Environment of Scotland* (NCC, Edinburgh, 1989).

24 Alderman, D. J. 'Fisheries chemotherapy', in Muir, J. F. and Roberts, R. J. (eds) *Recent Advances in Aquaculture* (London, 1988) pp. 1–62.

25 Austin, B. 'The control of bacterial fish diseases by antimicrobial compounds', in Woodbine, M. *Antimicrobials* (1984) pp. 255–68.

Index